£19.25

ARISTOTLE ON THE GENERATION OF ANIMALS

A Philosophical Study

Johannes Morsink

UNIVERSITY
PRESS OF
AMERICA

University Press of America, Inc.™
P.O. Box 19101, Washington, D.C. 20036

Printed in the United States of America

Library of Congress Cataloging in Publication Data

Morsink, Johannes.
Aristotle On the generation of animals.

Includes bibliographical references and index.
1. Aristotle. On the generation of animals.
2. Reproduction. 3. Reproduction--Philosophy.
I. Title.
QL41.A717M67 1982 591.1'6 81-43685
ISBN 0-8191-2606-3
ISBN 0-8191-2607-1 (pbk.)

TO MY PARENTS

PREFACE

Until very recently, the Generation of Animals was one of Aristotle's most neglected treatises. Even now, the only full-length commentary is the one by Michael of Ephesus, written in the sixth century.[1] Since then, most of the work done has taken the form of translations with notes and brief comments. At the second Aristotle Symposium in 1960, scholars met in Louvain to discuss the topic "Aristote et les problèmes de méthode."[2] Papers were presented on Aristotle's method of argumentation in the various treatises including some of the biological ones, but the Generation of Animals was omitted. Most of the participants did not make dialectic an integral and formal part of what they saw as Aristotle's scientific method. At the third Aristotle Symposium in 1968, scholars discussed Aristotle's Topics and his science of dialectic.[3] Again, the possible link between science and dialectic was played down, for no paper was presented on what Aristotle himself says is the primary purpose of dialectic, namely, the establishment of the archai (principles) of the various sciences. The papers of these two symposia give the impression that there was for Aristotle little or no connection between science and dialectic. A shift is taking place in that some scholars are now seeking a more unified picture of Aristotle's philosophy of science.[4] A rapprochement between Aristotle's views of science and of dialectic will have to be an important ingredient of this more comprehensive interpretation. This book is part of this new trend, for in it I argue that Aristotle used the dialectical method to establish the archai of biology; he used dialectical procedures to establish scientific principles. Although Anthony Preus has travelled some of the same terrain in his excellent paper "Science and Philosophy in Aristotle's Generation of Animals,"[5] this dialectical framework places some of the interpretations I share with him in a very different light. By interpreting Aristotle as being in dialectical opposition to the school of Hippocrates, I investigate and elaborate upon an episode in the history of science. I have tried to give an overview of the argumentative structure of the whole treatise, connecting the various books and passages with one dialectical thread.

In his Generation of Animals, Aristotle makes a distinction between form and matter in order to answer what, in his day, was an open question: Is there a difference between the male and female contributions to the reproductive process? Aristotle came to believe that the male supplies the form and the female the mere matter of offspring. This apparent biological sexism detracts from Aristotle's reputation as a great scientist.[6] A detailed textual and philosophical analysis of his treatise, however, shows that Aristotle wrote much of his argument in answer to two Hippocratic treatises. The author of *On Semen* and *On the Nature of the Child* assumed that like moves toward like and that the more specific pangenesis theory explains the phenomena of reproduction and inheritance. Aristotle's treatise, in turn, can be read as a sustained argument with this Hippocratic author. Aristotle did not think that the facts disclose in a simple, inductive manner the theories that explain them. As a practicing scientist, he had a historical and dialectical perspective on his work. Scientific progress for him was very much a matter of offering conjectures and refutations. In Book One, he points out the weaknesses of the pangenesis theory and asserts that his own form-matter hypothesis explains more. In Book Two, he corrects the presocratic maxim that like goes to like through an explanation of development based on a distinction between the potential and the actual. The dispute ends by Book Four, for Aristotle amends the form-matter hypothesis through principles of dominance and relapse; thus, he removes the apparent conflict between the hypothesis and those instances where the offspring resembles the female parent or ancestors on either side. Throughout the defense, Aristotle stays within the bounds of accepted scientific practice. When judged by the usual criteria of relevance, scope, and depth, Aristotle's use of the form-matter distinction in biology appears, rather than an expression of blind, unscientific sexism, to be a forward step in the history of science.

A greatly condensed version of the argument here presented was published under the title "Was Aristotle's Biology Sexist?" in the *Journal of the History of Biology*, vol. 12, no. 1 (Spring 1979), pp. 83-112. With minor revisions, chapter one appeared under the title "The Mandate of *Topics* I, 2," in *Apeiron*, vol. XVI (December 1982), pp. 71-97. I thank the editors of both journals for permission to use some of this material.

I am eager to acknowledge my indebtedness to Drew University for the secretarial and editorial services of Ms. Judith Kemp. It is also a great pleasure to thank my family for its support; my wife, Deborah, for her encouragement in word and deed, and my sons, Erik and Alexander, for their patience.

Johannes Morsink

Drew University
Madison, New Jersey

FOOTNOTES

[1]Michael of Ephesus (Pseudo Philoponus), Commentaria in Libros de generatione animalium, ed. W. Hayduck, Commentaria in Aristotetelem Graeca, vol. XIV, 3, Berlin, 1901. The most recent comments are those of D. M. Balme in his Aristotle's De partibus Animalium I and De Generatione Animalium I (with passages from II, 1-3) (Oxford: Clarendon Press, 1972).

[2]Published under that title by Publications Universitaires, Louvain, 1961.

[3]Aristotle on Dialectic, The Topics, ed. G. E. L. Owen (Oxford: Clarendon Press, 1968).

[4]See articles by Weil and Barnes in Articles on Aristotle, Volume I Science, ed. T. Barnes, et.al.(London: Gerald Duckworth and Col, Ltd., 1975).

[5]Journal of the History of Biology, Vol. 3, no. 1 (Spring 1970), pp. 1-52. (I encourage the reader to consult this paper for more details.)

[6]See Maryanne Cline Horowitz's paper "Aristotle and Woman," Journal of the History of Biology, vol. 9, no. 2 (Fall 1976), pp. 183-213.

TABLE OF CONTENTS

Page

PREFACE v

Chapter

I. SCIENCE, DIALECTIC, AND G. E. L. OWEN . . . 1

A. The Prevailing Opinion 1
B. A Difference on a Point of Method . . . 5
C. Owen's Solution 12
D. Critique of Owen 18
E. Conclusion 29

II. DIALECTIC AND HYLOMORPHISM 37

A. The Dialectical Character of Research . 37
B. Pangenesis or Hylomorphism 43
C. A Genuine Hypothesis 50

III. FOR AND AGAINST PANGENESIS 67

A. For Pangenesis 67
B. Against Pangenesis 73

IV. THE POTENTIAL-ACTUAL DISTINCTION 87

A. The Distinction Introduced 87
B. The "That From Which" Defended 100
C. The "That By Which" Explained 106
D. The Acquisition of Souls 111
E. Conclusion 119

V. THE THEORY OF INHERITANCE 127

A. The Unfinished Argument of Book I . . . 127
B. The Theory of Inheritance 131
C. Conclusion 141

VI. EVALUATION 147

A. Preliminary Assessment 147
B. A Bold Conjecture 149
C. The Vacuity Charge 155
D. The Charge of Vitalism 161
E. _Dunameis_ as Theoretical Entities 166
F. Conclusion 169

CONCLUSION 177

INDEX 181

CHAPTER ONE

SCIENCE, DIALECTIC, AND G. E. L. OWEN

In the first section of this chapter, I argue that most scholars (wrongly) believe that Aristotle saw little or no connection between science and dialectic. This prevailing negative view feeds on the last chapter of the *Posterior Analytics*. Aristotle says there that opinions, because of their fallibility, cannot play a role in the establishment of the *archai* or fundamental principles of a science. Such *archai*, he says, must be derived from sensory experience. But in the *Topics*, Aristotle seems to reverse himself and is far more positive about the relationship between science and dialectic. Here he claims that the *archai* of each of the sciences must be established dialectically, that is, by starting from the received opinions on the matters under discussion. Thus, while in the *Analytics* opinion is barred from the structure of scientific knowledge, in the *Topics* it lies at the foundation of that structure.

The difference between these two approaches to the *archai* of the sciences is great enough as to raise the question of whether and how they can be reconciled into one coherent picture of Aristotle's philosophy of science. In Sections C and D, I discuss and reject Professor G. E. L. Owen's proposed solution of this discrepancy in the relation of science and dialectic. Owen believes the discrepancy is only apparent and not real because the two disciplines have different and complementary tasks. As he sees it, dialectic is to establish the conceptual framework within which normal science is to be practiced. I reject this conceptual-empirical dichotomy because it does not fit the texts. Even in those discussions which Owen says are dialectical Aristotle appeals to experience.

We are, therefore, left with the recommendation that we study the argumentative structure of Aristotle's scientific treatises with this question of the relationship between science and dialectic in mind.

A. The Prevailing Opinion

Among Aristotle scholars, the prevailing opinion is that for Aristotle science and dialectic were not closely connected. This opinion can be found to reign in two groups of commentators, those who have commented on his scientific method and those who have commented on the nature or purpose of dialectic. I shall discuss a few representative figures of each of these groups.

An important member of the first group is G. H. Lewes, who in 1864 published the book Aristotle: A Chapter from the History of Science.[1] Chapter VI is entitled "Aristotle's Method." Lewes discusses here what he regards as the basic methodology which Aristotle followed as a scientist and he calls Aristotle

> the father of Inductive Philosophy, since he first announced its leading principles; and announced them with a completeness and precision not surpassed by Bacon himself.[2]

Thus Aristotle is presented as a forerunner of Francis Bacon. In his discussion Lewes stresses Aristotle's reliance "on Experience and Induction: the one furnishing the particular facts, from which the other found a pathway to general facts--or laws."[3] I am not now concerned with evaluating just how much of Bacon can be found in Aristotle, but I do wish to point out the fact that in this chapter on Aristotle's scientific method not a word is said about dialectic. Thus the reader is forced to conclude that dialectic plays no role at all in Aristotle's scientific method.

A much later and more contemporary member of this same group is Marjorie Grene, who in 1963 published her Portrait of Aristotle.[4] Ms. Grene differs from Lewes in that she does briefly consider the possibility that dialectical procedures play a role in the processes of science. In her chapter "The Tools of Philosophy" she discusses the source of our knowledge of the first principles or archai of the sciences. She briefly entertains the possibility that such principles are confirmed by dialectic. Says Professor Grene,

> Such confirmation could all be dialectical, argument tied only to common opinion and common experience. But then it could achieve only the lesser accuracy of a practical or a productive science, not the exactitude of theory.[5]

Again the implication of a remark like this is that there can for Aristotle be no connection between science and dialectic. Dialectic, she later adds, unfortunately without explanation, can only help in "the introductions to science, in the interstices of scientific argument."[6] The reader is led to believe that such introductions do not form an integral part of the scientific enterprise as Aristotle saw it. Even so, what is the extent of the help which dialectic gives to science? This we are not told. The usefulness of dialectic for science, we may safely infer, is of no great weight. Says Professor Grene,

> The heart of Aristotle's method lies in the great bridge from the definition of substance at the opening of the Organon to the account of real definition and finally of nous in Book II of the Posterior Analytics.[7]

Although Professor Grene has ever so briefly considered the possibility that science and dialectic are intimately related, the outcome of this consideration is clearly a negative one.

There are of course numerous other scholars who have commented on Aristotle's scientific method.[8] As a rule they do not pursue the connection between science and dialectic. And when they do point out the important fact that Aristotle was in the habit of surveying the views of his predecessors, they do not make this fact an integral part of Aristotle's method as a scientist. I think we can conclude that there is a tendency among these scholars to believe that for Aristotle dialectic had very little or nothing to do with science. I do not wish to go beyond claiming that there is this tendency, partly because I lay no claim to having read all that has been said on this matter and partly because there is at least one scholar, J. H. Randall, who did not succumb to the tendency.[9]

The other group consists of those scholars who have commented on the nature or purpose of dialectic. In the second chapter of the first book of the Topics, Aristotle states the purpose of dialectic. I shall discuss this passage in some detail in the next section. Its main points, however, must be noted now. Aristotle says here that there are two ways in which dialectic is helpful for the philosophic sciences. First, there is a general use by which dialectic helps us see two sides of any issue placed before us, and thus enables us the better to arrive at the truth. Second, there is a far more specific use of dialectic by which it is used to establish the basic principles or archai of each of the sciences. It is my claim that commentators have stressed the general use of dialectic at the expense of the particular use, thus lessening the tie between dialectic and science. I shall give two examples out of this group.

In 1935, L. M. Regis published a book entirely devoted to a discussion of Aristotle's opinion on opinion, L'Opinion selon Aristote.[10] When he comes to discuss the purpose of dialectic and its connection with science Regis says this:

> Indeed, once we have disposed of all the accessory material which makes the Topics a sum total of "lieux," a "vade-mecum" for argumentative minds, and once we have extracted the extremely ingenious procedure which is hidden under this mass of disparate propositions, then the Topics reveals to us the value of its methodological clarity and precision as well as its investigative power. This value and power are used by philosophers for their express purpose, the service of science and truth.[11]

This service performed by dialectic for science is never pursued in terms of the specific purpose of the establishment of the archai of the sciences. At the end of the above quote Regis footnotes the following remark, "the proper (idion) use of the dialectic is to be a way (hodos) towards science."[12] Clearly in the text the general use of dialectic for science is stressed and elaborated upon. And this happens at the expense of the proper and peculiar use of dialectic, comment upon which is relegated to a footnote. And even then Regis makes no mention of what that proper use of dialectic precisely is.

More recently, Professor Friedrich Solmsen spoke in the same vein as did Professor Regis. In commenting on the purpose of dialectic he said:

> To be sure, dialectic has (as we can here learn) a certain value also for the pursuit of serious philosophical studies, but where this value is defined the phrasing leaves no doubt that dialectic is one thing and philosophical subjects another.[13]

And he too delegated comment upon the more specific use of dialectic to a footnote.[14] Again, the general use of dialectic was stressed at the expense of what Aristotle called the peculiar use of dialectic.

By ignoring Aristotle's claim that dialectic is to be used for the establishment of the archai of each of the sciences these scholars adopted an interpretation of

Aristotle which would almost naturally lead them to believe that there is for Aristotle only a vague and extremely general connection between science and dialectic. In the case of this group too I need to mention an exception. J. M. Le Blond saw that there is for Aristotle an important connection between dialectic and science, but he never pursued this potentially fruitful insight at any great length.[15]

Even though, as I indicated, there are exceptions to the general rule, it seems clear that the prevailing opinion about what Aristotle saw as the relationship between science and dialectic is a negative one. According to most scholars there simply is not much of a connection. In the next section I shall point out what I take to be the main source of this negative view. I shall also argue that this view needs to be balanced by the more positive view found in the Topics.

B. A Difference on a Point of Method

There is in Aristotle's preaching on method a difference on how the first principles or archai of each of the sciences are to be established. Whereas in the Posterior Analytics he gives to experience, induction, and nous the mandate to establish the archai of the sciences, in the Topics this same mandate is given to dialectic. The first mandate contributes to the prevailing negative opinion about dialectic and science, while the second mandate provides a corrective note. I shall discuss these mandates in this order, ending with the positive correction.

I have already drawn attention to Professor Grene's belief that the heart of Aristotle's scientific method is to be found in the Organon, especially in Chapter 19 of Book II of the Posterior Analytics. This chapter is indeed a common rallying point for those who knowingly or unknowingly hold to the negative view about the connection between science and dialectic. This often quoted chapter fosters the negative view in the following way.

In Posterior Analytics II, 19, Aristotle finally tackles the question of how we acquire knowledge of the principles of archai of the demonstrative syllogisms he has been discussing. There can be, he says, "no knowledge through demonstration unless we know the immediate first principles" (protas archas--99b2). These first principles cannot themselves be established, however, by demonstration, for the basic premises of any supposed first demonstration would themselves have to be demonstrated and this would go on *ad infinitum* (I, 2, 72b5-25). Hence, Aristotle is forced to cast around for another way by which such principles are known, a way that does not involve an infinite regress of demonstrations.

He finds that way in the processes of perception and memory, which together yield the constituents of experience, experience being a series of memories of perceptions of the same kind of object (100a4-6). This experience in turn activates within the soul a faculty by which we then apprehend the universal characteristic which all the particulars we perceived had in common (100a7). Thus we are led from a perception of Callias and from a series of perceptions and consequent memories of other men to the apprehension of the universal man and possibly to the more ultimate universal animal (100b1-3). In this kind of an inductive way, Aristotle says, we establish the first principles (*prota*) of the sciences (100b4).

The faculty by which we apprehend these *archai* is *nous*, *nous* being the only faculty which is more accurate and less liable to error than the scientific knowledge built upon the *archai* supplied by *nous* (100b8). Aristotle rejects other possible candidates such as opinion and reasoning because these are liable to error. His point is that the *archai* of the demonstrative syllogisms must be known to be true with even greater certainty than the demonstrations based upon them (100b5-10). The main reason why *doxa* or opinion is rejected as the fountainhead of scientific knowledge is that opinions can be mistaken (100b7). Such mistaken opinions would undercut the certainty and infallibility of the scientific enterprise. Hence he concludes that "since besides scientific knowledge we have no other infallible faculty, *nous* must be the starting point of scientific knowledge" (100b14).

The connection I made here between *nous* and sensory experience is not atypical for Aristotle. We must not view *nous* as some sort of intellectual intuition, called upon in times of trouble--in this case to avoid the infinite regress of demonstrations. Professor Lesher has pointed out that already in the pre-Socratic writings and occasionally in Plato *nous* is seen "to act in close cooperation with the senses and to possess knowledge through their operation rather than in spite of them."[16] Aristotle often makes the same connection, the last chapter of the *Posterior Analytics* being a case in point. It is the function of *nous* generally to grasp universals after a repeated sensory perception. In this case, it just so happens that these universals are the _archai_ of the various sciences. But the underlying role of _nous_ responding to an inductive progression of perception and then grasping a universal remains the same.

In this discussion of how the _archai_ of the sciences are to be established, not a word is said about dialectic, and opinions, which are the starting point for dialectic, are explicitly barred from playing a role in the acquisition of scientific knowledge. If dialectic plays no role in scientific syllogisms nor in the establishment of the _archai_ of such syllogisms then, so the adherents of the prevailing opinion could safely argue, there is for Aristotle no important connection between dialectic and science. The negative view then finds a powerful defense in this mandate of the _Posterior Analytics_.

At first glance it would seem that in this respect the _Topics_ is no different from the _Posterior Analytics_, for in the first chapter the gap between dialectic and science is as wide as ever. In this chapter Aristotle compares and contrasts two kinds of reasoning, scientific reasoning and dialectical reasoning. The former kind of syllogism proceeds from premises which are true and primary (_alethōn_, _protōn_--100a23). The latter kind proceeds not from what is known to be true and primary but from generally accepted opinions or _endoxa_ (100b18). Of the _archai_, which form the basis of a scientific syllogism,

> it is not necessary to ask the question 'why,' but each of the _archai_ commands belief of itself. _Endoxa_, however, are opinions that seem to be true to all or to the majority or to the wise . . . (100b19-22).

The difference between the _archai_ and the _endoxa_ is that the former are self-evident truths which do not stand in need of further confirmation depending upon the truth of yet other propositions, while the _endoxa_ are not known to be true and do stand in need of further confirmation. _Endoxa_ only state what _seems_ to be the case, not what _is known to be_ the case. While scientific reasoning takes its departure from true propositions, dialectical reasoning cannot match the certitude of that foundation and must be content to start with what may or may not be true, i.e., with what seems to be the case. It would seem then that in the _Topics_ also Aristotle excludes opinions, even the generally accepted ones, from playing a role in the processes of science. Further reading, however, ought to convince us differently.

In the second chapter of Book I of the _Topics_, Aristotle discusses the uses of dialectic. From what he says there it is clear that he allows opinions, or at least the generally accepted ones, to play a very important role in the acquisition of scientific knowledge. The passage to which I already alluded in my discussion of the prevailing opinion is this:

> Dialectic is useful with respect to the philosophic sciences, for, being able to raise questions on both sides, we shall more easily discern truth and falsehood on each point. Further, it is useful with respect to the first [_archai_] concerning each science; for it is impossible on the basis of the proper principles (_oikeiōn archōn_) to discuss these themselves, since _the archai_ are the first of all [principles] and _it is_ necessary to arrive at them through the accepted opinions (_endoxōn_) concerning each issue. This is the _peculiar_ and most proper use of dialectic, for being investigatory in character it provides the way to the _archai_ of all methods of inquiry. (_Topics_, I, 2, 101a35-64)

In this passage, Aristotle gives two reasons why dialectic is useful for science. First, dialectic has a general usefulness in that it helps us see both sides of an issue and thus makes us better equipped to discern the truth or falsehood of a certain matter. Second, dialectic is useful with respect to the _prōta_ or _archai_, as they are variously referred to, of each of the sciences. It is important to note that this second use is the peculiar and most proper use of dialectic. Clearly, Aristotle thinks that there is a very close connection between science and dialectic.

Now these _archai_ are the very same _archai_ which Aristotle in the previous chapter had contrasted with _endoxa_; they are the _archai_ of the scientific syllogisms. Each science has its own peculiar principles (_oikeiōn archōn_). Since these principles are the most basic of all the premises within a certain science they cannot be established within the framework of that science. That is to say, the basic premises of a science cannot be established in the same manner as the less basic ones, which are deduced from them. The *archai* referred to in the first chapter are not identical with the *endoxa*, which are the starting point of dialectical reasoning, but apparently such *endoxa* can, by various dialectical procedures, be turned into or contribute to the establishment of the *archai* of a science. What at one time only *seems* to be *true* can, with the help of dialectic, be turned into a known truth or falsehood and can in the former case be used as the *archē* for a scientific syllogism. Thus, although there is in chapter one of the *Topics* a large gap between *endoxa* and *archai*, Aristotle goes on to point out that the former provide the dialectical basis for the latter. The mandate in the *Topics* assigns an essential role to the very opinions rejected in the *Posterior Analytics*.

The difference between the two treatises is basically this. In the *Posterior Analytics* Aristotle says that the *archai* of the sciences are to be established inductively and as a result of repeated sensory perceptions. The road leads straight from sensation to science. Fallible opinion is explicitly barred from playing a role in the acquisition of scientific knowledge. In the *Topics*, however, Aristotle recommends a very different procedure for getting at the *archai* of a science. He there recommends a dialectical procedure which takes its departure from the received opinions (*endoxa*) in any given field of investigation. Now opinions are made to play a key role and the growth of scientific knowledge is presented as a matter of conjectures and refutations. The reference to Popper is deliberate, for underlying these two different mandates seem to be two different views on the acquisition and growth of scientific knowledge. The pure inductivist and Baconian Aristotle in the *Analytics* meets us as a contemporary anti-inductivist in the *Topics*. Naturally, I am reading between the lines when I say this. But there is no escaping the fact that there are two different mandates here which, even if my schematization of the difference is overdone, is still a matter of puzzlement.

Ross has sought to reconcile the difference by pointing out that there are two modes of dialectic. He had in mind chapter twelve of the first book of the *Topics* where Aristotle says that induction (*epagogē*) in one kind of dialectical procedure and reasoning (*sullogismos*) is another kind (105a11-14). Says Ross:

> The statement that the first principles of science are approached by way of

> dialectic is nowhere brought into relation with the other statement that they are approached by induction; but we must remember that induction is one of the two modes of argument proper to dialectic.[17]

The suggested solution is that there is no discrepancy between dialectic and induction, both being roads to the first principles of science, because induction is one of the modes of dialectic--an express lane perhaps.

I do not think this suggestion works. Ross' reconciliation of the two mandates requires two suppositions that work in opposite directions. First, Ross supposes, *pace* Owen,[18] that the processes of induction of the two treatises are one and the same. This means that induction in the *Topics* does not start from the *endoxa* (while reasoning does), because otherwise the two processes would not be the same, opinion being barred in the *Analytics*. I think Ross is right about induction in the *Topics* not starting from the *endoxa*. What the two modes of dialectic have in common is that in both cases one is arguing with an opponent whom one seeks to convince. So, if Aristotle gave the mandate in the *Topics* to the inductive mode of dialectic there would be no problem.

But here I disagree with Ross. Clearly the mandate spelled out in chapter two of the *Topics* refers to the reasoning mode of dialectic, i.e., to that mode which starts from the *endoxa*. The mandate is not given to the inductive mode which Aristotle does not mention till ten chapters later. However, we wish to describe the process of arriving at the *archai* as spelled out in chapter two, it is clear that it starts from the received opinion. Either induction, too, starts from the *endoxa* (which suggestion I reject), in which case we have two processes of induction, the *Analytics*, one which bars opinions and the *Topics*, one which welcomes them. So the difference is still with us. Or, induction does not start from *endoxa*, but the mandate is given to the reasoning mode which does. Either way, Ross' suggestion fails us.

Still Ross has drawn our attention to an interesting chapter. Immediately after Aristotle says that there are two modes of dialectic, he explains the difference as follows:

> Induction is more convincing and clearer and more easily grasped by the senses, and

> common to many people, but reasoning is more cogent and more effective with argumentative opponents. (I, 12, 105a17-19)

Could it be that the difference between the two mandates is simply one of context and not as substantial as I have made it seem? Note that what Aristotle says here about induction fits the linkage between *nous* and sense perception of which I spoke earlier. Perhaps one's opponent determines what mode of dialectic one should use--induction with the ordinary man who lives on the level of his sensory experiences and reasoning with a more argumentative opponent. The latter presumably has his own interpretation of the data and cannot be convinced simply by being told what the facts are.

Tempting though this solution is, it is not very satisfactory. For one thing, the difference between the two modes as here described is not necessarily connected with the matter of establishing the *archai* of a science. There is no indication that this division of modes occurs *within* the mandate of chapter two. I think it does not. Furthermore, the language in both the *Analytics* and the *Topics* is too strong to let the difference boil down to one of context. In the *Analytics* Aristotle considered the role of opinion and explicitly barred such a role because it would undercut the entire structure of scientific knowledge. And in the *Topics* Aristotle recommends the dialectical (reasoning) mode for arriving at the *archai* not for heuristic or pedagogical reasons but because it is the only way to do it. It is necessary (*anagkē*-101b2) to do it this way; these basic *archai* or principles cannot be deduced from other more basic ones. Such *archai* must be arrived at by starting from the received opinions (*dia de tōn peri endoxōn anagkē dielthein*-101b2). We are not given our choice of which mode to use depending upon the situation.

The upshot is that there is a difference in Aristotle's preaching on method which makes it difficult to arrive at one coherent picture of how Aristotle viewed the relationship between science and dialectic. I know of no clear way of reconciling these two different approaches, at least not on a theoretical level. Le Blond is right when he calls for a close scrutiny of the various treatises with this problem in mind:

> It is difficult to resolve this controversy about the relationship between dialectic and science without a serious inquiry into Aristotle's practice itself. But,

> what we already know about the object and procedures of this discipline is enough to show that we do not have the right to set up as a principle its absolute separation with science.[19]

This is precisely what has happened in the prevailing, negative opinion. The absolute separation of science and dialectic has been set up almost as a principle of interpretation. I have shown how this negative view must be balanced by what Aristotle says in the _Topics_. The details of this more positive link between the disciplines depends, of course, on what a detailed study of the argumentative structure of the various treatises reveals.

Professor Owen has already made this sort of an investigation and has, as a result, proposed a way of reconciling the two mandates.

C. _Owen's Solution_

Owen believes that, fundamentally, Aristotle's view of the nature of science is of one piece and has no hidden cracks, as I tried to show in the preceding section.[20] He does think that there is an apparent discrepancy between the two mandates I described, but he feels that any such discrepancy as there may seem to be can be satisfactorily resolved. In his solution of the problem, Owen uses the conceptual-empirical contrast to explain the difference between dialectic and science proper. He attempts to show that this very difference between conceptual problems (dialectic) and empirical problems (science proper) is the reason that these two disciplines complement each other. Dialectic, Owen claims, establishes the conceptual framework within which science proper is to take place.

Owen does not explicitly speak of two mandates, as I did, but he does believe that there is in Aristotle (what later turns out to be an apparent) disagreement on "the means by which the principles of science are reached."[21] This disagreement is reflected in Aristotle's ambiguous use of the words _phainomena_ and _aporia_.

Owen does not make much of _Posterior Analytics_ II, 19. Instead he appeals to a passage from the _Prior Analytics_.

> It falls to experience to provide the principles of any subject. In astronomy,

> for instance, it was astronomical experience that provided the principles of the science, for it was only when the phainomena were adequately grasped that the proofs in astronomy were discovered. And the same is true of any art or science whatever.[22]

Phainomena here means "empirical observations"[23] or "the observed facts." Owen does not use this phrase, but it clearly fits the Baconian interpretation he places on the passage I just quoted: "The phainomena must be collected as a prelude to finding the theory which explains them."[24] According to this mandate the archai of a science, which are "the set of special hypotheses and definitions, from which the [scientific] deductions proceed,"[25] are derived from empirical observations. This description fits well with what I have called the mandate of the Posterior Analytics, though I did not base my argument on the meaning of phainomena, which term does not occur in that chapter.

Together with the Baconian sense of phainomena goes the use of aporia in the sense of "empirical problems." Whenever Aristotle takes his departure from experience he is worried about empirical problems. Thus in places where Aristotle is practicing the Baconian method of arriving at the archai, such as in biology and meteorology, the problems "tend to be questions of empirical fact or of the explanations of such facts or the problem of squaring a recalcitrant fact with an empirical hypothesis."[26]

For the second non-Baconian sense of phainomena Owen uses a passage from the Nicomachean Ethics:

> But in other contexts similarly concerned with methods of inquiry phainomena has another sense. In the Nichomachean Ethics Aristotle prefaces his discussion of incontinence with the words: 'Here as in other cases we must set down ("tithentas ta phainomena") the phainomena and begin by considering the difficulties, and go on to vindicate if possible all the common conceptions (endoxa) about these states of mind, or at any rate most of them and the most important.'[27]

Here the word phainomena does not mean "empirical observations," but rather "received opinions." Owen argues against Ross, who translated this occurrence of phainomena with "observed facts." He believes that Ross' translation "was evidently designed to bring Aristotle's programme into conformity with such passages as those already cited [from the Prior Analytics]"[28] The implication of Owen's objection (which I discuss in detail in the next section) is that in the preface to the discussion of incontinence phainomena means the same thing as endoxa and should be translated as such. It refers to the "common conceptions"[29] people have about incontinence. Owen believes that this non-Baconian and dialectical sense of phainomena is indicative of a different approach to the archai of a science and that Aristotle practices this method in his discussion of incontinence in the Ethics and in the one of place in the Physics.

Together with this dialectical sense of phainomena (supposedly) goes an equally different sense of aporia. When Aristotle takes his departure from the endoxa, he is not worried about empirical problems, but rather about philosophical problems or conceptual puzzles. In the discussion of incontinence, for instance, Aristotle is said to struggle with "logical or philosophical puzzles generated as such puzzles have been at all times, by exploiting some of the things commonly said."[30] Similarly, the problems of the Physics are "not questions of empirical facts, but conceptual puzzles."[31] In short, when Aristotle uses the dialectical method and starts from what is commonly said he is worried about entirely different kinds of problems than when he uses the Baconian method; the first deals with conceptual problems and the second with empirical ones.

Fortunately, says Owen, this discrepancy is only apparent, for even "if there is more than one use for the expression phainomena, the uses have a great deal in common."[32] Both the empirical observations and the received opinions must ultimately be checked against experience. Even the so-called observed facts must in the last analysis be tested against "the perceptual phainomenon that is reliable when it occurs."[33] The "facts" too must be tested to see whether they really are facts. And,

> In the same fashion the endoxa must pass the appropriate scrutiny, but in doing so they too become firm data... Endoxa also rest on experience, even if they misrepresent it.[34]

Since both the observations and the opinions must ultimately be tested against experience there is no discrepancy between the two methods or mandates.

It is important to note that Owen does not give up the difference between the two methods. He still believes that dialectic deals with conceptual problems and the Baconian method with empirical problems. Nevertheless, he affects a reconciliation of the two procedures by claiming that ultimately they both must be tested against experience. This seems paradoxical to say the least. For if dialectic deals with conceptual problems and establishes analytic truths, how can it at the same time have recourse to experience?

Owen's description of the dialectical method is in a large part the source of the problem. Note how he glosses the following translation:

> If so we shall have to say that the man of practical wisdom is incontinent, but no one would say this. (Not that it happens to be false, but that given the established use of the words it is absurd.)[35]

And again: "But we say (i.e., it is a common form of words) that some men are incontinent, without further qualification."[36] In a dialectical discussion Aristotle is said to start from the established ways of speaking. The implication is that he uses those ways of speaking as finally authoritative in resolving philosophical disputes. In yet another place, Owen reports Aristotle as ready to argue <u>a priori</u> that there is no use for a certain expression.[37] This <u>a priori</u> argument is presented as being an appeal to the established ways of speaking. Owen clearly seems to move in the direction of analytic-synthetic distinction when he refers to the "the analytic certainty" of certain "established forms of language" to which Aristotle is said to appeal.[38] These descriptions do indeed, as Terry Penner has suggested, lead us to suppose that Owen has the analytic-synthetic distinction in mind in his differentiation of the two methods as dealing with conceptual and empirical problems respectively.

But if this distinction does underlie Owen's discussion and interpretation of Aristotle, it can hardly be the traditional understanding of the distinction. If he has the traditional view of the analytic-synthetic distinction in mind, then his apparent discrepancy will turn out to be a real one. For how can there be common ground between a method which exclusively establishes analytic truths and one which exclusively establishes synthetic truths. Or, how can a method which takes no recourse to experience have anything in common with a method which does? Understood in the conventional sense the use of this distinction to explain the difference between the two methods would lead to an unbridgable gap and to a real instead of an apparent discrepancy. But Owen does believe in a reconciliation between the two points of view. So he cannot have a traditional understanding of the analytic-synthetic distinction in mind.

One passage contains a clue to a non-traditional interpretation of this distinction. Commenting on the use of dialectic in the Physics Owen says:

> The analyses of the Physics proper are preliminary to other more empirical enquiries and consequently must be justified in the last resort by their success in making sense of the observations to which they are applied. But this is not to say (and it does not commit Aristotle to supposing) that in the Physics proper the analyses either start from or are closely controlled by our inspections of the world.[39]

The context shows that these remarks are applicable to the dialectical method generally as well as to the analyses of the Physics. The words "preliminary," "more empirical," and "not . . . closely" point to a non-traditional interpretation of the conceptual (analytic)--empirical (synthetic) contrast Owen is using. Various such readings suggest themselves.

It could be that Owen has Quine's[40] views in mind and interprets Aristotle as follows. Dialectic establishes the propositions close to and right at the center of the conceptual scheme, while the Baconian method establishes the "more empirical" truths at the periphery. In this way the two methods complement one another because the one lays down the initial underpinnings of a structure to be further worked out by the other. Dialectic is preliminary to science proper. The former as well as the latter need to be tested against experience, but we presumably would be more ready and willing to give up the truths established by science proper than those established by dialectic.

But should the need arise, those dialectical truths, too, can be overthrown and replaced. A fatal objection to this interpretation is that whereas Owen does seem to accept an almost absolute distinction between conceptual and empirical truths and thus seems to accept at least some version of the analytic-synthetic distinction, Quine rejects the distinction in *toto*.

Nor can we read Putnam's[41] views into this remark by Owen about Aristotle's *Physics*. Though Putnam accepts the analytic-synthetic distinction as part of ordinary language, he hastens to add that the distinction cuts no philosophic ice. But Owen does think the distinction has philosophic merit and uses it to explain two very different tasks in the scientific enterprise. We have seen that Owen uses the terms "analytic" and *a priori* to describe the "framework principles" which *dialectic* establishes. Putnam, however, explicitly argues that these principles are not analytic truths, not even in some non-traditional or watered down sense.

The conception of the analytic and the *a priori* which seems to fit Owen's interpretation of Aristotle the best is the pragmatic one. In *Mind and the World Order*[42] C. I. Lewis argues that the *a priori* categories of the mind need not be fixed, but can be changed should the development of a certain science demand such a change. Though purely stipulative, these *a priori* categories or analytic truths are not arbitrary. Their adoption depends in large part on whether they make better sense of our total experience and agreed upon facts than do other possible categories and definitions. There is an empirical component built into the *a priori* element of knowledge. Of the various *a priori* schemes the mind can adopt it will pick the one that is the simplest and fits our experience the best. Though the difference between the analytic and synthetic has been maintained and is as absolute as it ever was, the adoption and acceptance of certain analytic truths into a certain conceptual scheme is itself not an *a priori* matter, but very much a question of what our experience demands of our concepts.

If Owen has this conception of analyticity in mind, then my objection is that he has not gone far enough. Having dealt at length with the apparent character of the discrepancy he should have made more of the fact that it is only apparent. That is to say, he ignored the empirical component of the *a priori* and hence arrived at a description of dialectic which does not fit the text of Aristotle. Owen and I are agreed that dialectic has as task the establishment of the *archai* or framework principles of the various sciences. But while he thinks that that method is almost exclusively concerned with conceptual problems and has no direct recourse to experience, I believe that the dialectical method does have direct recourse to experience. And this is so even if we hazard the interpretation that Aristotle held something like a pragmatic conception of the *a priori* and, furthermore, that he tied that conception closely to the task of dialectic. And I am not sure about this last point. It seems to me that even if dialectic establishes the *a priori* categories of a science, it also has the task of establishing a fundamental scientific hypothesis in a given field. Take, for instance, the form-matter hypothesis, i.e., Aristotle's belief that the male supplies the form and the female the matter of the offspring. Is that an *a priori* category or simply a crucially important scientific hypothesis offered within the framework of certain other even broader categories? The line is difficult to draw.

For that reason I refrain from claiming that Aristotle gave dialectic the task of establishing the *a priori* categories of the sciences. Perhaps Aristotle saw this as part of the task. But even if he did, Owen's absolute distinction between conceptual and empirical problems is still out of place. Even if the concepts of incontinence and of place are such *a priori* concepts, I argue in the next section that in the discussions of these concepts Aristotle does have direct recourse to experience. It may well be that in dialectical discussions Aristotle does at times concern himself with matters of definition, but such concern does not preclude his having recourse to experience, which is what Owen thinks. The conceptual-empirical distinction is therefore not very helpful in explaining the difference between the two methods, for in dialectical discussions also Aristotle has recourse to experience.

D. Critique of Owen

I agree with Owen that the discussions of incontinence and of place are dialectical discussions in which Aristotle seeks to establish certain basic principles or *archai*. But I do not agree with him in his claim that in these dialectical discussions Aristotle takes no (direct) recourse to experience. I shall argue this shortly.

But first I want to comment at length on Owen's claim that Aristotle uses phainomena in an ambiguous manner. This is not so. The word phainomena can be consistently translated with "what appears to be the case" or "the apparent facts."[43] This translation fits perfectly well both in the alleged Baconian and in the dialectical contexts of the word. Owen twice refers the reader to De Coelo III, 7, 306a16-17. Aristotle there designates "the perceptual phainomenon that is reliable when it occurs as the touchstone for the correctness of physical theories."[44] Owen's own translation of these lines inadvertently tells us what the real meaning of phainomenon is. It would be redundant to render phainomenon in this passage with "observed fact," for if an observed fact is what it is said to be then there is no need to add "which is reliable when it occurs." This addition is part of the meaning of the word "fact." But when we translate phainomenon with "apparent fact" then the addition does make sense and no redundancy results. Such a fact can turn out to be a fact, pure and simple, or it can turn out to be merely an apparent fact and not a real one. The touchstone of natural philosophy is the apparent fact which has stood up under further investigation.

No oddity or strangeness occurs when this translation is used in the Baconian contexts of phainomena. It makes perfectly good sense to read Aristotle as saying in the Prior Analytics that "it was only when the apparent facts were adequately grasped that the proofs in astronomy were discovered" (I, 30, 46a18).

The same holds for the dialectical contexts. Owen translates the relevant lines of the preface to the discussion on incontinence as follows:

> Here as in other cases we must set down the phainomena and begin by considering the difficulties, and so go on to vindicate if possible all the common conceptions (endoxa) about these states of mind, or at any rate most of them and the most important.[45]

He objects to Ross' translation of phainomena with "observed facts" and, as I mentioned in Section C, believes that the word here means roughly the same thing as endoxa and that both words in this preface should be translated with something like "received opinions." He objects to Ross' reading:

> For in the first place, what Aristotle proceeds to set out are not the observed facts but the *endoxa*, the common conceptions on the subject (as the collocation of *phainomena* and *endoxa* in the preface would lead us to expect.)[46]

I do not think there is this collocation in the preface at all. Even if it were true that the list is entirely made up of what people believe about the incontinent man that does not mean that *phainomena* here means the same thing as *endoxa* used two lines later. It makes perfectly good sense to suppose that Aristotle wants to start by laying down the apparent facts and that his way of finding out what those facts are is to see what people generally say about incontinence. The *endoxa* are the evidential support for the claim that the *phainomena* regarding incontinence are such and such. This relation by no means suggests, as Owen seems to think, that the two words mean the same thing. Of course not. The apparent facts refer to the state of the incontinent man while the received opinions are opinions about that man.

Owen gives a second quite complicated reason why we should translate *phainomena* in the preface with "received opinions."

> And, secondly, after this preliminary survey Aristotle turns to Socrates' claim that those who act against their own conviction of what is best do so in ignorance, and says that this is plainly in conflict with the *phainomena*. But he [Aristotle] does not mean that, as Ross translates it, 'the view plainly contradicts the observed facts.' For he remarks later that his own conclusion about incontinence seems to coincide with what Socrates wanted to maintain.... So Socrates' claim conflicts not with the facts but with what would commonly be said on the subject, and Aristotle does not undertake to save every thing that is commonly said.[47]

Owen's argument is this: 1) If in his earlier reference to Socrates (1145b28) Aristotle says that Socrates' theory conflicts with the facts, and if 2) at a later point (1147b15) Aristotle says that Socrates is right after all, then we must conclude that 3) Aristotle knowingly held a position which conflicts with the facts. But 4) clearly this is only the correct interpretation when all else fails. Therefore, 5) we must adopt the reading of _endoxa_ for _phainomena_ and so reconcile these two references to Socrates. In this case we find that at 1145b28 Aristotle disapproves of Socrates' rejection of one part of the _endoxa_ about incontinence, while at 1147b15 we find him in agreement with Socrates in rejecting another and different part of those _endoxa_. The translation of "observed facts" cannot be broken down in this way because presumably Aristotle would not want to reject even a part of the observed facts. Owen believes his reading avoids these difficulties.

The argument is ingenious, but it will not do. Though Owen's translation of _phainomena_ in the preface with "received opinions" would indeed reconcile the two references to Socrates, the same is true for the more faithful translation of "apparent facts." On this second interpretation Aristotle first says that Socrates' theory conflicts with part of the apparent facts and he later says Socrates was right in taking exception to another part of the apparent facts. There is nothing odd about this reading, for it is no surprise that after closer scrutiny part of the apparent facts turn out to be only apparent. Presumably Aristotle would go along with Socrates in rejecting that part.

I conclude that Owen is incorrect when he claims that Aristotle used _phainomena_ in an ambiguous manner between "observed facts" and "received opinions." Consequently he cannot base his explanation of the difference between dialectic and science (proper) on this point of translation.

This does not mean that Owen is therefore also wrong in his more substantial claim that in his discussions of incontinence and of place Aristotle has no (direct) recourse to experience. While the translation of "apparent facts" creates the initial presumption that Aristotle is interested in the way the world is, it is still possible that in the course of the dissussion he does in fact take no recourse to experiene and only appeals to established ways of speaking about incontinence and place. In the remainder of this section, I shall argue that in both these discussions Aristotle does have (direct) recourse to experience. The conclusion will be that Owen's characterization of dialectic as mostly and primarily a conceptual enterprise does not fit the texts of his own examples and must, therefore, be rejected. And with it goes his solution of the discrepancy between the two mandates.

One of the key statements Owen makes in the defense of his case is the following:

> He [Aristotle] is anxious, unlike Socrates, to have use for the expression 'knowing what is right, but doing what is wrong,' but he is ready to show *a priori* that there is no use for the expression 'doing what is wrong in the full knowledge of what is right *in the given circumstances*.'[48]

Owen refers here to two expressions regarding incontinence. First, there is the general statement that incontinence is "knowing what is right, but doing what is wrong." Second, there is the more specific statement that it is "doing what is wrong in the full knowledge of what is right in the given circumstances." While Socrates rejected both statements, Aristotle accepts the general one but rejects the specific one. The point to which I wish to draw attention is Owen's claim that Aristotle rejects the second more specific statement (where the italicized phrase is added) on *a priori* grounds. I grant that most of the discussion of incontinence turns on the issue of whether or not the incontinent man does what is wrong while he knows it is the wrong thing to do *in his specific circumstances*. While I agree that Aristotle rejects this description of the incontinent man, I have two objections to Owen's claim that he does so on *a priori* grounds.

First, I believe Owen is inconsistent. The claim that Aristotle offers an *a priori* argument against the specific expression conflicts with Owen's earlier point about the translation of *phainomena*. He sought to reconcile the two references to Socrates by arguing that at 1145b28 Aristotle disagrees with Socrates in the latter's rejection of one part of the *endoxa*. This part must be the general statement that incontinence is "knowing what is right, but doing what is wrong." Later, at 1147b15, Aristotle agrees with Socrates when the latter rejects another part of the *endoxa*. This part must be the specific expression with the italicized phrase, which Owen says in the above quote Aristotle is ready to argue against. This means that Owen claims that Aristotle argues *a priori* against a part of the *endoxa* or the received opinions.

But how can this be? Owen himself described an a priori argument as one that appeals to the established ways of speaking. (See Section C.) And are not the endoxa themselves the established ways of speaking on a subject? If so, Owen presents Aristotle as arguing on the basis of established forms of speaking against (other?) established forms of speaking. Owen's interpretation ends up in confusion. He admits that the specific expression (which Aristotle rejects) is part of the endoxa. How, then, can he go on to claim that Aristotle argues against this expression on a priori grounds, i.e., by appealing to the endoxa. Clearly, something is wrong. So, my first objection to Owen's claim that Aristotle is ready to argue a certain point a priori is that it conflicts with Owen's own defense of his translation of phainomena with "received opinions." For one cannot argue from the endoxa against the endoxa. The only option left for Owen is to give up the translation argument and to remove the specific expression from the status of being an endoxon. Having done that he can then claim that Aristotle rejects this expression (which now is not claimed to be part of the established ways of speaking) on a priori grounds.

My second objection is that, even with this internal conflict removed, Owen's argument still is defective.

Aristotle gives four arguments against the specific expression and if Owen is right he has no recourse to experience in any of the four. I think he does have such recourse and that his arguments are a posteriori. Aristotle's use of deinos in the preface to the discussion is I think an indicator of what we can expect in the four arguments. So I begin with the preface.

One of the received opinions Aristotle reports on in the preface is that "the incontinent man out of passion seems to do things he knows to be evil" (1145b13). Let us call this proposition (P). Socrates did not think (P) was true,

> for it would be strange (deinon) in the case of someone who had knowledge that something else would overpower it and drag it about like a slave. (1145b24)

Aristotle's use of deinos here is important, for he uses it again in the first of the four arguments to be discussed. Socrates disagrees with (P) not because he thinks that anyone who uttered (P) would be guilty of conceptual confusion, but because he thinks that the state of affairs described by (P) would be physically or physiologically impossible. It would be strange or surprising if a state of knowledge could be blocked out and overpowered by something else, i.e., it is hard to see how this could be true.

Aristotle has reported Socrates as raising an empirical problem and he intends to respond in part at least with an empirical investigation:

> Now this theory [of Socrates] is clearly in conflict with what seems to be the case (phainomena) and we must make an investigation of this state of mind (pathos), whether it is due to ignorance and what kind of ignorance is involved. (1145b28-29)

Aristotle, we must note, does not call for an investigation of the established ways of speaking about incontinence, which he might have done had he felt Socrates' charge on the conceptual level, but for an investigation of a state of mind. He intends to go beyond the endoxa about incontinence, directly to the state of mind itself.

This is his plan of action, but can he implement it? In defending the received opinion that there is such a thing as a state of incontinence Aristotle faces the same sort of problems as anyone else who tries to defend the existence of mental states against the charge that there are no such things. Negatively, all one can do is refute the arguments of one's opponent. Positively, about the only response can be to assume that there is such a state of mind and to describe it as carefully as one can, hoping one's opponent will see the light. This Aristotle does in the first two arguments. Also, one can, especially if one's opponent does not deny the existence of mental states generally (and Socrates does not), try to explain how the particular state of mind is caused. This Aristotle does in the third and especially in the fourth argument.

In the first argument against Socrates (1146b31-35), Aristotle differentiates between two kinds of knowledge, knowledge one has but is not actively aware of, and knowledge one has and of which one is also actively aware. Aristotle observes that it would be strange (deinon) if someone did wrong while he was actively aware that it was wrong, but it would not be strange if he had the knowledge but was not actively aware of it. Owen thinks that the use of deinos here is an indication of Aristotle's readiness to argue a priori. It would be absurd to say of a man who did the wrong thing that he was actively aware that it was wrong. To say this would display conceptual confusion on the part of the speaker.

I think Owen has read too much between the lines. First, Aristotle does not say why it is deinon. Second, in the absence of such an explanation it is better to interpret deinon in light of the earlier occurrence discussed above. In that context deinon meant surprising. There is no warrant here for introducing the notion of conceptual confusion.

In the second argument (1147a1-10) Aristotle distinguishes between the knowledge of a universal premise and the knowledge of a particular premise. He then goes on to say that it would not be extraordinary (atopon) if a certain incontinent man knew the universal premise that dry food is bad for man. But it would be amazing (thaumaston) if he exercised the knowledge of the particular premise that this food is dry food. Again Owen takes Aristotle's use of atopos and thaumastos as an indication that he is ready to argue a priori. He takes Aristotle to argue that, if someone said that an incontinent man knew that this food is dry food, in such a case the speaker is guilty of conceptual confusion. There is, however, no indication in the text that this is what Aristotle means. Instead, the more plausible interpretation is to say that Aristotle argues that it would be surprising if the incontinent man knew the particular premise. Why this is so will become clearer from the next two arguments.

In the third argument (1147a11-24), which Owen omits in his reference, the nature of the dispute becomes quite clear. Aristotle compares the state of incontinence to the state of being asleep or drunk. In the last two states the body is somehow affected, so that even if a drunken man, for instance, rattled off the verses of Empedocles, he does not really know or understand them. Such a man is not conscious of them in the way a sober man would be. This difference marks the line between real knowledge and knowledge which is ignorance in disguise. Comparing incontinent persons to those who are drunk or asleep, Aristotle says, "it is clear that we must say (lekteon) that those who are incontinent are affected similarly." We are told to speak in a certain way because physiologically incontinence is like being asleep or drunk. This imperative, based as it is on an appeal to the facts, gives added impetus to the interpretation I have given of deinon, atopon, and thaumaston. If the way the world is determines how we ought to speak, it probably also determines how we ought not to speak.

The fourth argument (1147a24-b19) to which Owen refers starts out with a statement which makes it very unlikely that Aristotle is here arguing a priori. "Further, one could study the cause of incontinence scientifically (physikos) as follows." (1147a25) Ross' translation reads "Again, we may also view the cause as follows with reference to the facts of human nature."[49] This translation fits perfectly well with what Owen himself has said about Aristotle's use of physikos:

> The [Baconian] method is expressly associated with physikē and physikos and from the stock examples in these contexts--astronomy--it seems clear that the phainomena in question are empirical observations.[50]

Owen is faced with the predicament that in the opening statement of an argument in which he says Aristotle is ready to argue a priori, Aristotle uses a word which Owen himself says is the mark of the rival Baconian method.

The predicament is worsened once we note that in 1147b6-9 Aristotle openly addresses himself to the empirical problem of how the state of incontinence is dissolved and again compares it to the state of sleep or drunkenness. Finally, at the end of the argument (1147b15-18) Aristotle says he agrees with Socrates in that the knowledge of a particular premise concerning what is sensible can be dragged about. From these final remarks, it is clear that the disagreement has not been about what expressions do or do not make sense.

I conclude that in his discussion of incontinence Aristotle does have recourse to experience.

Owen's second example, the discussion of place, can be dealt with more briefly. Owen thinks that

> if the Physics is to be described as setting out from a survey of the phainomena it is plainly this second [endoxa] sense of the word that is more appropriate.[51]

Again, Owen argues that the four arguments Aristotle advances on behalf of the existence of place are appeals to established ways of speaking rather than to experience or empirical observations:

> [This discussion] opens with four arguments for the existence of place of which the first states what dokei (it appeals to the established ways of talking about physical replacement), the third states what certain theorists legousi, the fourth quotes what Hesiod and the majority nomidzousi and the remaining [second] one relies on the doctrine of natural places which is later taken as an endoxon.[52]

Owen takes Aristotle's use of dokei, legousi, nomidzousi, and endoxon to be indicative of his interest in and of his appeal to the way people talk about place. The implication is that Aristotle regards these ways of speaking as authoritative in deciding upon the reality or unreality of place.

The first argument goes like this:

> The existence of place is held to be obvious from the fact of mutual replacement. Where water now is, there in turn, when the water has gone out as from a vessel, air is present. When therefore another body occupies this same place, the place is thought to be different from all the bodies which come to be in it and replace one another. What now contains air formerly contained water, so that clearly the place or space into which and out of which they passed was something different from both.[53] (208b1-8)

Naturally Aristotle is interested in the claim that people judge place to have an independent existence. But the real weight of the argument comes from the observations that support this judgment. We judge place to have an independent existence because we see that air is now in the same place water was. The phenomenon of physical replacement, the seeing that one thing is where another used to be, is evidence that place is independent of its contents.

The second argument relies on the theory of natural places, according to which each of the elements--earth, air, fire, and water--has its own place to which it naturally goes.

> To us they [natural places] are not always the same but change with the direction in which we are turned: That is why the same thing may be both right and left, up and down, before and behind. But in nature each is distinct, taken apart by itself. It is not every chance direction which us 'up,' but where fire and what is light are carried; similarly, too, 'down' is not any chance direction but where what has weight and what is made of earth are carried--the implication being that these places do not differ merely in relative position, but also as possessing distinct potencies (208b15-22).[54]

Aristotle defines "up" and "down" not in terms of the way we might be tempted to talk on the basis of our own relative position, but in terms of what the natural places are to which the elements go. He appeals to empirical observation of where the elements naturally go to decide what "up" and "down" is.

Of this second argument Owen says that Aristotle refers to it later as an endoxon, but this word is not used. Aristotle does say at 210b33-34 that from now on in the discussion he shall take for granted that the characteristics which are judged (dokei) to belong to place do in fact belong to it. Two of these characteristics are the independent and objective nature of place, which were established in the opening two arguments I discussed above. The same advice is repeated at 211a9, when Aristotle states as an objective of the further investigation that "the characteristics which are judged to belong to place (dokounta) do in fact belong." Owen is right when he interprets these dokounta as the judgments and definitions of people, but he overlooks or ignores the fact that in the preceding discussions Aristotle appealed to certain empirical observations as relevant to the acceptance of these opinions. Consequently, in the rest of these discussions these judgments are taken for granted and so are the relevant empirical observations.

The other two arguments mentioned by Owen (208b26-27 and 208b31-33) are mere appendices to these first two. What authority they do have, they receive because their conclusion agrees with the one of the first two arguments. Aristotle reports these opinions with approval because they confirm the observations already reported.

I conclude that in his discussion of place Aristotle does have direct recourse to experience. Assuming this to be a dialectical discussion and one that took its departure from the endoxa, it clearly is not a discussion in which conceptual problems and puzzles are the sole focus of attention.

E. Conclusion

The conclusion of this chapter is both constructive and destructive. On the constructive side I argue (in Sections A and B) that the view of the relationship between science and dialectic, given to us in the Posterior Analytics, especially in the last chapter, needs to be balanced by what we read in the Topics. While in the former place Aristotle bars opinions and hence dialectic from playing a role in the processes of science, in the latter he gives to dialectical reasoning the mandate to establish the archai of each of the sciences. Since such reasoning starts from the received opinions on a certain topic, we are faced with a discrepancy in Aristotle's preaching on a point of method.

I myself know of no clear way to fit these two different mandates together into a coherent picture of how Aristotle viewed the relationship between science and dialectic.[55] I suggest that we need to study the argumentative structure of the various treatises so that with this added detailed information we can fill out the picture and perhaps resolve the discrepancy. The very fact though that Aristotle does give the dialectical mandate in the Topics suggests that the prevailing negative view is too negative and needs to be amended. We must look for places where Aristotle consciously practices the mandate of the Topics and we must not, as many scholars seem to do, start out with the assumption that science and dialectic have very little or nothing in common. This is my positive contribution.

Professor Owen is an exception to this prevailing negative opinion. I explained in the third section his interpretation of two discussions in which Aristotle is engaged in the dialectical establishment of certain archai. On the basis of an analysis of Aristotle's discussions of incontinènce and of place, Owen concludes that dialectic and science are two complementary disciplines for Aristotle. The former provides the analytic truths that are to guide the research of the latter.

Tempting as it is (and here I become destructive), we must reject Owen's solution of the discrepancy. He supposes that in his dialectical discussions Aristotle has no direct recourse to experience and is worried not about empirical but about conceptual problems. In the fourth section I have shown, however, that Aristotle does have direct recourse to experience in his discussions of incontinence and of place. Consequently, Owen's use of the conceptual-empirical distinction as a way to describe the difference between the dialectical and Baconian methods is too neat and too precise. It does not fit the text. Aristotle does not seem to have operated with a clear line between conceptual and empirical problems, for even in dialectical discussions experience plays an important role.

We are therefore left with Le Blond's and my suggestion that the argumentative structures of the various treatises be studied with this question of dialectic, vis-á-vis, science in mind. In the next chapter, I begin my study of the argumentative structure of the Generation of Animals. I shall defend the claim that Aristotle uses the dialectical method to establish the archai of biology. In the Generation of Animals Aristotle practices what he preaches in the Topics.

FOOTNOTES

[1]G. H. Lewes, Aristotle: A Chapter from the History of Science (London: Smith Elder and Company, 1864).

[2]Lewes, p. 108.

[3]Lewes, p. 109.

[4]Marjorie Grene, A Portrait of Aristotle (Chicago: University of Chicago Press, 1963).

[5]Grene, p. 110.

[6]Grene, p. 112.

[7]Grene, p. 112.

[8]Ingmar Düring, Aristoteles (Heidelberg: Carl Winter Universitätsverlag), "Der 'Sekretar' der Natur," pp. 514-54; Werner Jaeger, Aristotle (Oxford: Clarendon Press, 1948), Chapter 13, "The Organisation of Research"; G. E. R. Lloyd, Aristotle: The Structure and Growth of his Thought (Cambridge University Press, 1968), Chapter 4, "The Philosopher of Nature," and 5, "The Founder of Systematic Research." From the title of Chapter 4, one can see that Lloyd does not stress the Baconian line as much as others, but still dialectic plays no role in his interpretation of Aristotle's concept of science. Also, T. E. Lones, Aristotle's Researches in Natural Science (London: West Newman & Co., 1912), Chapter II, "Aristotle's Method of Investigation in the Natural Sciences"; J. Needham, A History of Embryology (New York: Abelard-Schuman, 1959), Chapter I, "Embryology in Antiquity"; C. Singer, Greek Biology and Greek Medicine (Oxford: Clarendon Press, 1922), Section 2, "Aristotle," pp. 18-54; D'Arcy W. Thompson, On Aristotle as a Biologist (Oxford: Clarendon Press, 1913).

[9]J. H. Randall, Jr., Aristotle (New York:Columbia University Press, 1960), Chapter III, "Science as Right Talking: The Analysis of Discourse." Professor Randall is one of the few scholars who have discussed the importance of dialectic for an understanding of Aristotle's conception of science. See also note 55.

[10]L. M. Regis, L'Opinion selon Aristote (Ottawa: Publications de l'Institute d'Etudes Médiévales, 1935).

[11] Regis, p. 142. Unless otherwise stated, all translations are my own; for the ones from French, I am indebted to my wife, Deborah.

[12] Regis, p. 142, note (4).

[13] F. Solmsen, "Dialectic without the Forms," in Aristotle on Dialectic. The Topics, ed. G. E. L. Owen (Oxford: Clarendon Press, 1968), p. 53.

[14] Solmsen, p. 54, note (4).

[15] J. M. Le Blond, Logique et Méthode chez Aristote (Paris: Librairie J. Vrin, 1939), pp. 222-68. See also note 55.

[16] James H. Lesher, "The Meaning of nous in the Posterior Analytics," Phronesis, Vol. XVIII, No. 1-1973, p. 49.

[17] Sir David Ross, Aristotle (London: Methuen & Co., Ltd., 1966), p. 57.

[18] G. E. L. Owen, "Tithenai ta Phainomena," reprinted in Aristotle: A Collection of Critical Essays, ed. by J. M. E. Moravcsik (Garden City, N.Y.: Doubleday & Co., Inc., 1967), p. 171: "Epagōgē can be said to establish the principles of science by starting from the data of perception [here Owen footnotes Posterior Analytics II, 19]. Yet epagoge is named as one of the two cardinal methods of dialectic [here he footnotes Topics, I, 12] and as such must begin from the endoxa. . ."

Owen's paper was first published in Aristote et les problèmes de méthode (Louvain: Publications Universitaires, 1961). All my references are to the Moravcsik volume of critical essays.

[19] Le Blond, p. 45.

[20] Owen's paper entitled "Tithenai ta Phainomena" has two parts. My comments are exclusively based on the first part of Owen's paper. In the second part he points out that in the Physics Aristotle takes his departure from Plato's Parmenides rather than from the Timaeus. This thesis falls beyond the scope of the present discussion about the relationship between science and dialectic. Even if Owen is correct in the second part of his paper, he need not be right in what he says in the first part.

[21] Owen, p. 168.

[22] Owen's translation, p. 168.

[23] Owen, p. 169.

[24] Owen, p. 168.

[25] Owen, p. 168.

[26] Owen, p. 171.

[27] Owen, p. 169.

[28] Owen, pp. 169-170.

[29] Owen, p. 170.

[30] Owen, p. 172.

[31] Owen, p. 173.

[32] Owen, p. 174.

[33] This is Owen's translation of De Coelo III,7,306al

[34] Owen, p. 175.

[35] Owen's translation of N.E. VII, 3,1146a5-7;p. 172.

[36] Owen's translation of N.E. VII, 3,1146b4-5; p. 172.

[37] Owen, p. 170.

[38] Owen, p. 174.

[39] Owen, p. 176.

[40] Willard Van Orman Quine, "Two Dogmas of Empiricism in From A Logical Point of View (New York: Harper & Row, 1963). "The totality of our so-called knowledge of belief . . . is a man-made fabric which impinges on experience only along the edges . . . but the total field is so under-determined by its boundary conditions, experience that there is much latitude of choice as to what statements to reevaluate in the light of any single contrary experience. No particular experiences are linked with any particular statements in the interior of the field, except indirectly through considerations of equilibrium affecting the field as a whole." "If this view is right, it is misleading to speak of the empirical content of an

individual statement--especially if it is a statement at all remote from the experiential periphery of the field. Furthermore, it becomes folly to seek a boundary between synthetic statements, which hold contingently on experience, and analytic statements, which hold what may. Any statement can be held true come what may, if we make drastic enough adjustments elsewhere in the system" (pp. 42-43).

[41] Hilary Putnam, "The Analytic and the Synthetic" in Minnesota Studies in the Philosophy of Science, Vol. III, Herbert Feigl and Grover Maxwell (eds.), University of Minnesota Press, 1962. "It is not, of course, an accident that one will consistently be wrong if one attempts to employ the analytic-synthetic distinction in philosophy. 'Bachelor' may be synonymous with 'unmarried man' but that cuts no philosophic ice. . . It is the belief that there are synonymies and analyticities of a deeper nature, synonymies and analyticities that cannot be discovered by the lexicographer or the linguist but only by the philosopher that is incorrect" (p. 362). "But there are many, many principles--we might broadly classify them as 'framework principles'--which have the characteristic of being so central that they are employed as auxiliaries to make predictions in an overwhelming number of experiments, without themselves being jeopardized by any possible [add 'isolated' cf. p. 374] experimental results . . . . Any principle in our knowledge can be revised or abandoned for theoretical reasons unless it is really an analytic principle in the trivial sense in which 'All bachelors are unmarried' is an analytic principle"(p. 375).

[42] C. I. Lewis, Mind and the World Order (New York: Charles Scribner's Sons, 1929). "If the a priori is something made by mind, mind may also alter it. There will be no assurance that what is a priori will remain fixed and absolute throughout the history of the race or for the developing individual (p. 233). Consequently, the so-called "categories of physics" or the "categories of biology" do, for Lewis, serve as examples of a priori categories. "But when concepts are intended to be applied in experience, and a priori principles are to determine modes of classification and interpretation, the case is different [from the pure a priori principles in mathematics] . . . . But knowledge has a practical business to perform, the interests of action which it seeks to serve. The mode of the mind's activity answers to our need to understand, in the face of an experience always more or less baffling, and of our need to control" (pp. 237-38). "If the criteria of the real are a priori,

that is not to say that no conceivable character of experience would lead to the alteration of them."(p. 263)

[43] Here I am indebted to Terry Penner.

[44] Owen, p. 175, note 40.

[45] Owen, p. 169.

[46] Owen, p. 170.

[47] Owen, p. 170.

[48] Owen, p. 170.

[49] See _Basic Works of Aristotle_, ed. by Richard MacKeon (New York: Random House, 1941), p. 1041.

[50] Owen, pp. 168-69.

[51] Owen, p. 172.

[52] Owen, p. 172.

[53] R. P. Hardie's translation in _The Works of Aristotle_ (Oxford: Clarendon Press, 1930), Vol. II, _Physica_.

[54] Hardie's translation. I have added the emphasis.

[55] One possible way of bringing these two mandates closer together is to follow Weil and Barnes and to make a distinction between Aristotle's logic of presentation (or justification) and his logic of acquisition (or discovery). In that case the present study can be seen as an exploration of Aristotle's logic of discovery. I am sympathetic to this approach. However, since it is very much a matter of philosophical dispute as to whether or not the distinction between these two logics is a viable one, I have not cast my argument and exegesis in its terminology. See Barnes' article "Aristotle's Theory of Demonstration" and Weil's "The Place of Logic in Aristotle's Thought." Both can be found in _Articles on Aristotle Volume I Science_, edited by J. Barnes, M. Schofield, and R. Sorabji (London: Gerald Duckworth and Company Ltd., 1975). My sympathy is more explicit in another version of this chapter, published under the title "The Mandate of Topics I 2" in _Apeiron_.

CHAPTER II

DIALECTIC AND HYLOMORPHISM

I have contrasted the two mandates as being inductivist and anti-inductivist. Whereas in the *Posterior Analytics* Aristotle recommends a straight path from sensory experience to the principles of science, in the *Topics* I see him recommending the curvy path of conjectures and refutations. In section A of this chapter, I show that this second, dialectical mandate fits very well with Aristotle's own occasional reflections on how a scientist is to proceed and make progress in his research. Aristotle thought scientific research had to have a dialectical character and he took a problem-solving approach to his own work as a scientist.

In section B, I introduce the main participants of the dialectical dispute on which this study focusses. I argue that in the first book of the *G. of A.* (and by extension in the others as well) Aristotle is engaged with a certain Hippocratic author in a dialectical dispute about the *archai* of biology. While this author held the view that the semen contains portions from all the bodily parts of both parents and thus felt they made similar contributions to the generative process, Aristotle proposed his own form-matter hypothesis as a better way of solving the same problems. I lay out the skeleton of this dispute and show that it is indeed a dialectical dispute as defined in the *Topics*.

In Section C, I elaborate upon my use of the word "hypothesis" in describing Aristotle's position. I argue that Aristotle's belief that the male provides the form and the female the matter of the offspring was a genuine scientific hypothesis and not a dogmatic hangover from his Platonic days.

A. The Dialectical Character of Research

It is no coincidence that scattered throughout the Aristotelian corpus we find methodological remarks which in their content are remarkably similar to the mandate of the *Topics*. In that treatise, Aristotle preaches that a survey of the extant opinions on a problem or question is an integral part of the establishment of the *archai* of the relevant science. And this, as every student of Aristotle knows, is precisely what Aristotle practices when he commences so many of his scientific investigations with a survey of what the predecessors have said. This correlation between Aristotle's preaching and practice is no coincidence. At the beginning of many of his investigations Aristotle lays it down as an explicit methodological principle that the investigation must start with a survey of what others have said. I am thinking of the following passages: _Metaphysics_ A, 3, 983b1-7; 995a24-b4; L, 8, 1073 b13-16; M, 1, 108,6a1-20; _Physics_ IV, 4, 211 a7-11; _De Caelo_ I, 10, 279 b12; IV, 1, 308a6; _De Anima_ I, 2, 403b20-26; _Nicomachean Ethics_ VII, 1, 1145b2-6. In all these places, Aristotle makes more or less explicitly the same point, namely, that one must start an investigation with a survey of what others have said. The _endoxa_ are to be the starting point of the discussions. Judging from the great variety of treatises in this list, we must conclude that Aristotle often, and in a great variety of contexts, practiced the mandate of the _Topics_. One point of comparison between the mandate and these remarks needs especially to be brought forward.

Aristotle often says that the survey must not be simply one of what people have said but also one of the problems involved in what they have said, problems which must be solved in the subsequent discussions. The views of the predecessors are judged in part on how many problems they solve and in part on how many they raise. This point is clearly made in the _De Anima_ passage:

> In our inquiry about the soul we shall have to raise problems (_dia porountas_) for which we must find a solution, and as we progress we must take with us for comparison the opinions (_doxas_) of the predecessors on this subject in order that we may accept those that are well stated, and if something is not well stated we must be on our guard against it. (_De Anima_, I, 2, 403b 20-26)

As it turns out, many of the problems that need solving are ones that are raised by the formulations of the predecessors. Aristotle can then offer his own view of the soul as a superior one which does not raise the same sort of problems.[1] A survey of the views of the predecessors here as elsewhere ends up being a survey of the problems connected with those views. With this sort of beginning Aristotle prepares the way for a more ready acceptance of his own views.

This methodological emphasis on the solving of problems fits well with the context in which the mandate of the *Topics* is given. Not every proposition, we are told, is a dialectical one, but only those about which there can be disagreement. A dialectical problem is a question to which the answer is not immediately clear (104b14). It is "an investigation leading either to choice and avoidance or to truth and knowledge" (104 b1-2). The problem dealt with can be an ethical, logical, or physical one, including problems in biology (105 b15-29). Clearly, a dialectical discussion can deal with empirical problems, even though it need not, as in the case of logical problems. We learn from the *Topics*, as we did from the methodological remarks referred to above, that a serious dialectical discussion centers around a problem or problems to which the participants seek an answer. As Aristotle's practice reveals, more often than not the problems are created by the views of the predecessors. The *endoxa* are themselves problems or problematic questions.

We may now ask why Aristotle as a scientist took this problem-solving approach in his work. Ms. Mansion does not think that this question can be answered in a general way.[2] She thinks that in each of his surveys Aristotle has a different goal in mind and that no simple strategy underlies all the surveys equally. I believe that there is such a strategy. There is a passage in the *Metaphysics* which bears both on the preceding discussion and on the present question that arises out of it. I quote it at length:

> It is necessary with respect to the science being investigated that we first go over the questions that need to be raised. These are different things people have proposed concerning them, and also if something other than these happens to have been overlooked. It is for those who want their doubts resolved useful to thoroughly go through all the questions. For the later resolution of the doubts is the solutions of the questions raised earlier, but there are no solutions for those who are ignorant of the knot, but the perplexity of the mind makes this clear concerning the matter. For insofar as one is perplexed in that respect one has fallen into a situation very much like people who have been fettered. For in either case it is impossible to make

> progress. Hence it is necessary to first have surveyed all the difficulties, both for the reasons given (i.e. to remove perplexity) and also because those who make an investigation without first raising difficulties are like people who are ignorant of where they should go; and besides this, one does not know whether what one has searched for has been found or not. For to such a person the end is not clear, but it is clear to him who has first raised all the difficulties. Further he is necessarily a better judge who as happens in a lawsuit has heard all the opposing arguments. (Metaphysics B, 1, 995a24-b4)[3]

In this statement, Aristotle has gone beyond his usual methodological reflections. Not only does he say that the investigator should start by raising a series of problems, but he also says why this should be done. The reason is that the need to solve certain extant problems tells the researcher whether or not he has made progress. People who do not first specify certain problems to be solved lack direction in their research. There is no indication here that these remarks apply only to the principles of being that Aristotle is about to investigate. On the contrary, he has made a general methodological remark applicable also to all the other contexts where he says one ought to start with raising problems. The reason and underlying strategy is the same in all cases. Unsolved problems give direction to the investigation and serve as yardsticks for success and failure.

I conclude that for Aristotle, scientific research had to be dialectical in character. His own theories are often developed in a dialectical give and take with what he thought to be defective alternatives offered by his predecessors. It is against this general background that we must view Aristotle's dispute with his opponents. Before I discuss the details of that example, I shall first answer two objections which may be brought against the claim that Aristotle practiced the mandate of the Topics as often as I have said he did. One objection concerns the content of the discussions to which I have appealed and the other concerns their format.

As regards the content of the discussions, introduced by the methodological remarks listed above, it may be argued that since I have not said a word about which *archai* Aristotle seeks to establish in each discussion, I have no right to claim that he practices the mandate of the *Topics* in all these discussions. After all, that mandate concerns itself not just with dialectical discussions generally, but with the *archai* of the various sciences. So, what about these *archai* and what about the very concept of an *archē* about which I have said very little so far? My response to this objection is threefold. First, it is not very important for the defense of my thesis that I spell out in any detail what an *archē* is. Aristotle's own remark will suffice when he tells us that "it is common to all *archai* to be the first from which something either exists or comes to be or comes to be known" (*Metaphysics* V, 1, 1013a18-19). Evidently an *archē* of a certain field of investigation is either ontologically or epistemologically basic to that field. So, my thesis is that whatever these *archai* turn out to be in these various discussions, the stating and solving of certain specified problems is part and parcel of their being dialectically established. Since we have already seen a remarkable similarity between the methodological remarks which preface these discussions and the mandate of the *Topics*, it is quite likely that the parallel also holds in this respect of the *archai*. It is true, of course, that a full and complete defense of my thesis about the character of research requires a detailed investigation of all these discussions together with a statement of what *archai* are established in each case. Without that, the claim that Aristotle frequently practiced the mandate of the *Topics* has not been fully secured. Even so, quite a few parallels have been pointed out and in subsequent chapters I am going to discuss one example in detail.

Second, even if in a certain discussion the word *archē* were not used by Aristotle that does not necessarily mean that he cannot be practicing the mandate of the *Topics*. Owen and I agree, for instance, that Aristotle practices this mandate in his discussions of incontinence and of place. Yet, in neither of those discussions does Aristotle use the word *archē*. The absence of the word makes it simply more difficult to detect what *archai* Aristotle may be seeking to secure. Whatever in such a discussion would turn out to be epistemologically or ontologically prior, that would be the *archē*. Third, in some of the discussions to which I have appealed, Aristotle does use the word *archē*. In *De Caelo*, the four elements are referred to as the *archai* (III, 4, 302b4). In *Metaphysics* A, he refers to the four causes as *archai* (988a20), and in B he is going to uncover the archai of being (995b8).

As regards the content of the discussions to which I have appealed, Owen has raised and answered the following objection.[4] How can these discussions be labeled dialectical if they do not exhibit the yes and no or question and answer format so familiar to us from the Platonic dialogues and which is discussed by Aristotle himself in the Eighth Book of the Topics? Owen responds, and I concur, that there are, according to Aristotle, two kinds of dialectical propositions:

> For the man who is putting questions the dialectical protasis is an asking (an invitation to choose one) of a pair of contradictions. For the man who is reasoning to a conclusion it is an assumption of that which is accepted and accredited (Prior Analytics 24b10-13, Owen's translation).

The first kind of dialectical proposition clearly fits the Platonic dialogues and presumably the lost dialogues of Aristotle himself. The second kind of dialectical proposition fits the discussions we have been talking about, i.e., Aristotle's surveys of the views of his predecessors. These discussions start with an assumption of what is generally accepted (endoxou) and of what seems to be the case (phainomenou). It is explicitly stated that this too is a dialectical proposition even though the familiar yes or no format is absent.

To Owen's already adequate answer we may add that the question and answer format is not nearly so scarce in Aristotle's treatises as the objection supposes it to be. Very often Aristotle begins an investigation with a series of questions. For example, he opens the discussion which will serve as our example as follows:

> Therefore, this must be investigated whether all males discharge semen, or not all--and whether females contribute or not? (I, 17, 721a33-36, emphasis added)

This "whether--or not" format is typical of a dialectical problem (Topics I, 4, 101b30). Furthermore, the asking of questions and the stating of problems amount to the same thing. And Aristotle does raise a great many problems. For both the asking of questions and the raising of problems he regularly uses the verb aporeo. This verb is, for example, repeatedly used in the Second Book of The Generation of Animals.[5] Each time, a new question is asked or a new problem raised to which the subsequent discussion must find an answer.

I conclude that Aristotle regularly practiced the mandate of the Topics and that the stating and solving of certain specified problems is an integral part of the establishment of the archai of a certain science. Also, the stylistic differences between Plato's dialogues and Aristotle's treatises do not prevent us, not even in the eyes of Aristotle himself, from calling certain discussions in those treatises dialectical. More specifically, the example now to be discussed is not defective simply because it is not written in dialogue form.

B. Pangenesis or Hylomorphism

Throughout the G. of A., Aristotle refers to the male and the female as the archai of genesis.[6] Right after he has made the point, at the end of the first chapter of Book One, that the distinction between the sexes is almost universal, he begins chapter two with this remark:

> We may safely lay down the male and the female as the archas tes geneseos, the male as the one who has the principle (archēn) of movement and generation, and the female as the one who has the matter (I,2,716a4-7).

There are, of course, other archai involved in the generation of animals, such as the moon (777b24), and the hot and the cold (764a28). But Aristotle never refers to these factors as "the archai of genesis." The least Aristotle means by this is that the male and the female are the primary, initiating causal factors in the generation of animals.[7] This hypothesis is amply justified by the fact that in almost all cases of generation, a male and a female are involved. Aristotle spends the first sixteen chapters of Book One showing that this is so.

But, as the above quote shows, Aristotle did not think that the sexes were the same kind of archē or archai in the exact same sense of the word. Instead, he felt they made different contributions to the generative process and that each sex is an archē in his or her own particular way. Usually he refers to the contribution of the female as the matter of the offspring (e.g., I,2, 716a5; 9, 727b31, 21, 730a15 and 27; II,1, 732a6; 4, 738b21). The male's contribution is, however, variously referred to as the principle of movement (see above quote), which at 728b9 he uses interchangeably with eidos or simply as eidos (729b18), or as the shape (morphē - 729b7), or as the dunamis of the specific character (poian tina - 730a14). The variety of these descriptions of the male role raises the problem of what Aristotle means by eidos in Book One. To what does "form" in form-matter hypothesis refer?

We must resist the temptation of thinking we have to choose between two alternatives: eidos as species-form and eidos as individual-form. In the latter case, it refers to the shape of a particular male parent such as Coriscus. The form-matter hypothesis covers both.

At the very beginning of Book Two, Aristotle refers back to the discussions of Book One (chapters seventeen to the end). He says that only a species (eidos) and not its individual members can be eternal. Then, by way of further elaboration, he adds that the first moving cause to which the eidos belongs, being more divine, should exist separately from the female who supplies the matter (731b35-732a6). In this context, the male is clearly presented as the cause of the fact that the offspring is of the same (eternal) species as its parents.

But Aristotle also believed, and this comes through openly in Book IV, that the male parent ideally contributes his own individual shape to the offspring, his snub nose, short legs, and high forehead, etc. (Peck overlooks this when he argues that eidos--even in Book One--means "soul.")[8] The male supplies the poian tina and the morphē of the offspring. This sort of language suggests a more detailed form than the species kind of form requires. Furthermore, many of Aristotle's arguments against his opponent hinge on the fact that the eidos is a very particular form, such as a scar on one's left arm. "Form" then must be understood both in its broader species sense and its narrow individual sense. The male supplies both of these, while the female supplies the matter to be enformed. I shall refer to this thesis of Aristotle's as his form-matter hypothesis. (More on the word "hypothesis" in the last section.)

I am of the opinion that Aristotle attempts to establish this hypothesis dialectically. While from chapters one through sixteen Aristotle establishes *that* the male and the female are the *archai* of generation, from seventeen to the end he spells out *the sense in which* they are the *archai*. In contrast with the first half of the book, he makes no simple, straightforward appeal to the facts in the last part. Aristotle engages himself with a dialectical opponent who has different views on what the *archai* of biology are. He argues that an at-first-glance, viable alternative is full of defects and he offers his own hypothesis as a solution to the problems. He makes the implausibility of the views of his opponent an integral part of the defense of his own position.

Although Aristotle nowhere in the *G. of A.* mentions Hippocrates' name, it is quite clear that he is dialectically engaged with a member or members of that school. At 721b12, Aristotle introduces a pangenesis theory quite similar to the one Darwin was to propose much later.[9] Aristotle remarks that "since some [*tines*] say that it [*semen*] is drawn from the whole body, it must first be investigated how this works." My suggestion is that the *tines* here refers to a member or members of the school of Hippocrates who wrote the two little treatises *On Semen* and *On the Nature of the Child*.[10] There is a remarkable similarity between Aristotle's statement of the pangenesis theory and the contents of these Hippocratic treatises. It is not just that in *On Semen* we find a statement of the very theory Aristotle introduces and then attacks, but also that in these two treatises the theory is defended in a way that Aristotle says it can be defended. At this time, I shall only show the similarity between the two statements. The rest of the evidence will be presented in subsequent chapters.

In the third paragraph of *On Semen*, the Hippocratic author says this:

> I maintain that semen is separated from every part of the body (*apo pantou tou sōmatos*), from the solid parts as well as from the soft parts and from every fluid that is in the body.

Aristotle uses the same words when he says that some believe that the semen comes *apo pantou tou sōmatos* (721b13), and he uses this phrase regularly to describe the view of the opposition.[11]

The Hippocratic author further believed that (P) (= the pangenesis theory) held for both the male and the female.

> And in the semen itself there is some drawn from each part of the body (apo pantou tou sōmatos), both of the woman's and of the man's . . . (§8)

He makes the same point in the sixth paragraph:

> The following is furthermore to be noted. The semen of the woman is sometimes stronger and sometimes weaker; the same holds for the man. The result is that "a mixture is affected (misgetai) of that which comes from the man and that which comes from the woman.

Clearly this equality of the sexes conflicts with Aristotle's own form-matter hypothesis. If the male and the female both secrete semen from all their bodily parts, then they make similar contributions to the generative process, which is what Aristotle denies. Aristotle explicitly rejects the notion of a mixture of two equal contributions: oude mignumenon amphoin ginetai, hōsper tines phasin (I, 19,727b18), and hōs ou summignumenon (I, 21, 730a12). The two theories are incompatible; either the sexes make similar contributions or they do not (cf., I, 17, 721a33-36). Aristotle's interest in the truth or falsity of the pangenesis theory is well placed. If it is true that the semen comes from all the bodily parts of both male and female, then there is no room for his own hypothesis. Conversely, if (P) must be rejected, for whatever reasons, then there is room for the form-matter hypothesis. It would not conflict with another theory that was known to be true.

An inescapable question at this point is whether I am correct in identifying our Hippocratic author as Aristotle's main opponent. Could he not have been arguing with Democritus instead?[12] Or could he not have been arguing against both the school of Hippocrates and Democritus, as Lesky implies when she claims that the pangenesis theory of that medical school is rooted in the atomism of Democritus who also held a version of the theory.[13] Why identify one historical school and two of its (perhaps spurious)[14] treatises as Aristotle's main target? The answer is that the evidence is too compelling.

Several times in Book IV, Aristotle _seems_ to say that Democritus held (P) (1, 764 all, b 14, and 765a8). But a close reading of the text shows that these passages, to which both Peck[15] and Lesky[16] appeal, do not say that Democritus held (P). Aristotle nowhere explicitly nor implicitly says that Democritus believed that the semen of both male and female comes from all the bodily parts. Each time Aristotle explicitly mentions Democritus' name, it is in connection with the claim that the male is said to contribute some material (from his generative organ) to the offspring. Empedocles, too, is mentioned in this connection, and there is no evidence either that he held (P).

Furthermore, in chapter 18 of Book One, where he attacks his opponent with a series of fifteen arguments, Aristotle never mentions Democritus. Why, if he is willing to mention Democritus at various points in the treatise, should he drop his opponent's name where it counts most and substitute "some people say"? Also, none of the fragments in Diels (68a141 and 68b32-68a143 is the same as 764a6) actually contain the crucial phrase _apo pantou tou sōmatos_, which does repeatedly occur in _On Semen_.

Against the weak and speculative connections with the philosophy of Democritus stand the solid links between the _G. of A._ and the two Hippocratic treatises mentioned. I have already provided some of the evidence for this link and much more will be supplied in subsequent chapters. We can only view Democritus as Aristotle's major opponent if we are willing to make him the author of these two treatises which no one has suggested we do. The connection between these two treatises and what Aristotle says and does in the _G. of A._ is simply too tight to give up. But should it turn out that Democritus was their author, this will not hurt my thesis. I simply claim that Aristotle's major opponent was the author of these two treatises, whoever that was. By (tentatively) labeling these works as Hippocratic, I am following present opinion.[17] This opinion can be revised, but that is not the object of this study.

We are left then with our Hippocratic author as Aristotle's main opponent in a dispute about the _archai_ of biology. This historical connection is an interesting and important one. It reinforces the connection, left largely unexplored, between Aristotle's philosophical beliefs and procedures and the medical tradition from which he came. It also shows that in defending his hylomorphism, Aristotle is not first setting up a straw man readily to be knocked down. On the contrary, this is a genuine dispute with good arguments on both sides. Aristotle is coming to grips with views held by contemporaries of his who were members of the school of Hippocrates. It has been pointed out that some parts of the *Nicomachean Ethics* seem to reflect what were ongoing debates in the Academy.[18] The same may well be true of big sections of the *G. of A.* on which this study concentrates, such as the series of arguments against (P) in chapter eighteen. The discussion of semen in Book One could easily be an echo of a life and ongoing philosophical debate in the academy about the *archai* of biology.

Having introduced the antagonists in this dispute, it remains to show in the remainder of this section that the dispute is indeed a dialectical one.

In the *Topics*, Aristotle mentions some general features that dialectical disputes have in common.[19] I shall list the most important of those features and in each case argue that the particular feature characterizes the dispute between Aristotle and our Hippocratic author. The conclusion is that since this dispute has all these features, it is a dialectical dispute in the proper sense of that word.

First, a dialectical discussion starts from *endoxa*. These *endoxa* are the opinions acceptable to *all or* to the *many or* the wise, and of these either to all or to the most famous and distinguished of them (*Topics* 100b22-24). Aristotle starts his discussion of semen from what are the accepted opinions on the matter: "it seems that all things come to be from semen and the semen from the parents" (I, 17,721b17). More specifically, he starts from the thesis put forth by the school of Hippocrates to the effect that semen "comes from the whole of the body" (721b13). Throughout the entire treatise, Aristotle comes back to (P), often in order to point out the incompatibility between it and his own form-matter hypothesis.[20] Thus (P) is the *endoxon* from which the discussion starts.

Second, the problem of a dialectical dispute must not be trivial. The dispute must center on a substantial and controversial problem (*Topics* 104b1-17). Clearly, a discussion on the nature of semen and the basic principles of generation is not a trivial one.

The problems raised by (P) concerning the equal or unequal contributions the sexes make is an important problem. The arguments for and against (P) and the form-matter hypothesis clearly show that there is ample room for controversy in this dispute. Also, Aristotle clearly recognizes that the dialectical method might be used to deal with problems in biology (Topics 105b15).

Third, a dialectical discussion commonly starts by raising difficulties concerning the initial endoxon (Topics 101a34-36). These are raised in chapter eighteen where Aristotle catalogues fifteen objections against the pangenesis theory, which he has introduced and explained in chapter seventeen. I discuss these objections in the next chapter.

Fourth, a proper dialectical discussion must aim at establishing the archai of a certain science (Topics 101a36-b4). I have already referred to the numerous places where Aristotle speaks of the male and the female as the archai of genesis. The opening line of Book Two contains the most important passage:

> That the male and the female are the archai of genesis was said earlier and what the dunamis and the logos of the ousia of them is (II,1,731b18-20).

This is a backward reference to what Aristotle has already accomplished in the first Book. He must especially have in mind chapter seventeen through twenty-one of that book, for it is there that the dunamis of each of the sexes is established.[21] It is the being or ousia of the male to supply the form and of the female to supply the matter of the offspring. There is more to this discussion of semen than a first reading might suggest. In it, Aristotle is consciously and purposely engaged in the establishment of the archai of biology. He himself says so at the beginning of Book Two.

Fifth, as I argued in section A, Aristotle took a problem-solving approach to the nature of research. He generally presents his own view as an improvement over defective alternatives. In terms of our example, the form-matter hypothesis must then be offered as a solution to the difficulties of the pangenesis theory. And this is precisely what Aristotle does do. In his criticism of the pangenesis theory, he raises three kinds of problems. In one set of arguments, he tried to show that, contrary to what the defenders say, (P) does not help explain how it happens that children resemble their parents.[22] Aristotle presents his own view as an improvement over (P). I discuss Aristotle's response in chapters III and V. A second set of objections deals with the differentiation of the sexes and is discussed again by Aristotle in the beginning of Book IV.[23] I return to these issues in Chapter V. Finally, in a third set of arguments, Aristotle points out that the defenders of (P) cannot explain how a living, unified, and besouled animal comes from the embryo.[24] Aristotle uses the form-matter hypothesis together with the potential-actual distinction as part of his own explanation of how an animal develops. I discuss this in Chapter IV. I conclude that Aristotle offers the form-matter hypothesis as a solution to the difficulties raised by (P). In this, he follows his normal practice of the mandate of the Topics.

These five features are the most important characteristics of a dialectical discussion. The dispute between Aristotle and the defenders of (P) has all these features. I take it, therefore, as established that we have here before us an example of a dialectical discussion in the proper sense of that word, i.e., an example where Aristotle practices the mandate of the Topics.

C. A Genuine Hypothesis

In this last section, I redeem my promise of commenting on the hypothesis part of the form-matter hypothesis.

Biologists and philosophers alike have admired Aristotle's acumen as a scientist. Singer calls him "the greatest of all Greek men of science and perhaps the greatest of all men of science..."[25] Needham claims that with Aristotle,"general and comparative biology came on its own."[26] Very often the G. of A. especially is singled out for praise as a model scientific treatise. Needham (though he does not think the work is well organized, probably because he misses the dialectic of the argument) calls this work "the first great compendium of embryology ever written."[27] Thomas East Lones no doubt spoke for many a student of Aristotle when he said that the G. of A. "astonishes the reader by its deep philosophical reasoning and furnishes evidence of a powerful intellect grappling with obscure embryological problems."[28] It is generally recognized that Aristotle's discussion of development in Book Two has set the stage for the entire history of the debate between preformationists (like our Hippocratic author) and those who defend epigenesis (like Aristotle himself).[29]

Philosophers too have been full of praise. Ross, who almost never lapses into the language of praise and blame, notes with respect to the discussions of reproduction that "not only many of his [Aristotle's] most remarkable observations but some of his most penetrating discussions bear on this subject."[30] Marjorie Grene has taken the discussion of development, in which Aristotle makes use of the potential-actual distinction, as a model Aristotelian explanation.[31] Finally, to add a contemporary reference, Baruch Brody[32] recently argued that Aristotle's theory of scientific explanation is an improvement over the covering law model to which so many present-day scientists are committed. As the reader knows full well, I could extend this list of favorable reviews of the "Dean of Early Science"[33] and "the philosopher" endlessly.

Against this background of highest praise, it comes as a shock to find scholars maintain a deadly silence or turn quite critical when confronted with Aristotle's use of the form-matter hypothesis in biology. This is even more surprising in light of the fact that this hypothesis permeates the entire argumentative structure of the so-highly praised G. of A. Singer, who talks at length about Aristotle's observations in biology, does not mention the hypothesis in his history of embryology. But elsewhere,[34] he refers to it as a "peculiar view" which stands in the way of Aristotle's real contributions. Needham mentions the hypothesis in passing and gives the following explanation of why Aristotle may have held it. The belief that the mother has no real formative influence on the child was a general belief in Aristotle's day, probably borrowed from the Egyptians. In Aeschylus' Eumenides, for instance, Apollo defends Orestes against matricide with the following words: "The mother of what is called her child is no parent of it but nurse only of the young life that is sown in her."[35] This mythical belief may have been the rationale for putting all captured males to death so as not to pollute the conquering race. Needham presents Aristotle as standing squarely in the midst of this myth, unable to take critical distance from it. One of Aristotle's most fundamental scientific doctrines turns out to be the uncritical reflection of an ancient myth about mother earth.

Philosophers have been equally embarrassed by this apparent faux pas. Düring believes that in biology, Aristotle followed Plato's method more closely than Theophrastus' method. The result is that "everywhere in his biological writings we can see how he [Aristotle] adapts his observations or his knowledge about various things to certain sets of preconceived ideas."[36] The form-matter hypothesis--though not singled out--is, of course, a prime example of this approach. Lloyd praises Aristotle's "ingenuity in drawing out the implications of the available evidence," but then warns us that "we should observe how his [Aristotle's] own theoretical preconceptions influenced his solutions."[37] And he gives the form-matter hypothesis as an example. The praise that is given with one hand is taken away with the other.

The almost inescapable conclusion from the above remarks is that the greatest biologist and philosopher of nature ever to live made a colossal blunder which crippled the best scientific treatise he wrote. How can this be? Either the form-matter hypothesis is not a relapse into dogmatism or the uncritical acceptance of a myth, or Aristotle is not the great man of science that he is made out to be. Either the one or the other is the case, but not both, especially not since this hypothesis is crucial to everything Aristotle said about topics in biology. Either Aristotle does not deserve all the praise he has been given or the form-matter hypothesis must be brought in line with his reputation as a great scientist. I am going to do the latter.

Hypotheses are a necessary and indispensable part of the nature and progress of science. Views differ, of course, on the exact place of hypotheses in the structure and acquisition of scientific knowledge, but even an inductivist like John Stuart Mill believes that "nearly everything which is now theory was once hypothesis."[38] Mill believed that the use of hypotheses was a shortcut that bypassed the normal first inductive step toward scientific knowledge, thus leaving ground uncovered that later has to be retraced.[39] On the other end of the spectrum, we find a man like Popper claiming that the actual and only procedure of science is "to jump to conclusions [when faced with a problem situation] often after one single observation."[40] After that, one should test one's guess. And the bolder the guess or the conjecture, the better it is, because the very boldness of a conjecture leads to discoveries of interesting and relevant truths regardless of whether the conjecture turns out to be correct or incorrect.[41] In the last chapter of this study, when I come to evaluate the dialectical dispute between Aristotle and our Hippocratic author, I shall have something to say about this notion of boldness as it applies to Aristotle's form-matter hypothesis. For the moment, I want to elaborate a bit further on the idea of hypothesis itself. Since Popper will be heard of later, I now stay with Mill.

Mill defines an hypothesis as follows:

> An hypothesis being a mere supposition, there are no other limits to hypotheses, than those of the human imagination; we may, if we please, imagine by way of accounting for an effect, some cause of a kind utterly unknown and acting according to a law altogether fictitious. But as hypotheses of this sort would not have any of the plausibility belonging to those which ally themselves by analogy with known laws of nature, and besides would not supply the want which arbitrary hypotheses are generally invented to satisfy by enabling the imagination to represent to itself an obscure phenomenon in a familiar light, there is probably no hypothesis in the history of science in which both the agent itself and the law of its operation were fictitious.[42]

This is precisely the point of Aristotle's famous art-nature analogy; the known is called upon to help explain the unknown. Aristotle felt that distinctions which were useful in explaining the production of artifacts could also be of help in explaining the generation of animals:

> And from these instances (of carpentry) someone may learn how the male contributes to the process of generation. For not every male emits semen, and of the males that do, this semen is not a part of fetation as it develops, just like nothing passes from the carpenter to the wood, which is his material and no part of the art of carpentry is part

> of that which is being made, but the shape (morphē) and the form (eidos) come to be from that (carpenter) by means of movement in the material; his soul, in which the form and the knowledge are, moved his hands (or some other part) with a particular movement, a different movement for a different product, and the same movement for the same product, and his hands move the tools and the tools the material. Similarly, nature, in the male of animals that emit semen, uses the semen as an instrument and as what has movement in actuality just as in artifacts the tools are in movement, for the movement of the heart is present in them. As many then as emit semen contribute to generation in this manner. (G. of A., I, 22, 730b9-25)

This is not the legacy of Plato imposed upon the biological data, but a genuine hypothesis offered to solve a nasty problem.

The problem is that the pangenesis theory does not work. (I explain this in detail in the next chapter.) According to this theory, the male and the female make similar contributions to the reproductive process; they both secrete semen from all their bodily parts. This raises all sorts of questions: 1) why do the male and the female need each other? If the male (or the female) secretes semen from all his (or her) bodily party, why can't he (or she) produce an animal by himself (or herself)? 2) If both male and female secrete semen from all their bodily parts, why is the offspring not bisexual, having a portion of the reproductive organ of each parent? 3) Does not this theory conflict with the general principle that different tasks require different tools and, conversely, that different tools are indicators of different tasks? If the tasks of the male and the female are the same, as they are according to (P), how are we to account for the difference between the reproductive organs of the two sexes? This is just a sampling of the problems (P) raises.

I view Aristotle's form-matter hypothesis as a genuine attempt to solve all these problems with one bold conjecture. Could it not be the case that natural organisms are produced in roughly the same way as artifacts, where a certain form is imposed upon a certain piece of material? And if this were so, would it not explain and remove the difficulties raised by (P)? If the male's role is like that of a carpenter and the female's like that of the material to be shaped, these and other questions can be answered, provided the details can be worked out. There is no doubt that Aristotle himself viewed his hypothesis as an answer to the problems raised by the pangenesis theory and as an improvement. Twice during the discussion he makes the explicit methodological point that:

> Therefore it is part of the same argument whether both male and female emit seed or one only, and whether it comes from all of the body or not from all; for it is reasonable, if it does not come from all the body, we must first examine how this matter stands. (721b8-13)[43]

Aristotle has linked the implausibility and falsehood of (P) to the plausibility and possible truth of his own hypothesis. There is a logical interdependence here. Since the theories are incompatible and mutually exclusive, the downfall of the one advances the fortunes of the other. Since many of the reasons for rejecting (P) have to do with the problem of the differentiation of the sexes and the different roles the partners play in the reproductive act, these same reasons make the form-matter hypothesis all the more plausible, since that hypothesis acknowledges the very differences ignored by (P). Aristotle argues that, given a choice between (P) and his own hypothesis, the latter is far to be preferred over the former. It solves all the problems and raises no major new difficulties.

From this vantage point the seemingly erratic arguments of Book One, which bothered Needham as digressions from the main topic[44] and which I at first viewed as eristic overtones of the dispute,[45] fall into place. Aristotle does everything he legitimately can to bolster his own hypothesis, vis-a-vis, the pangenesis theory. To that end, he does the following things:

1) He argues that semen is a residue of final nourishment and not a colliquation. The latter is a breakdown of tissue and a sort of (un)natural by-product of the natural processes. Since the defenders of (P)

believe that the semen comes <u>from</u> all the bodily parts they must think it to be a disposable by-product (724 b36). But this is unlikely, argues Aristotle, because while a colliquation is contrary to nature (725a3) and runs about in the body freely (a35), semen clearly is a natural product and has ducts provided by nature for it to flow through and a natural product or organism comes to be from it (which incidentally is the reason why it cannot be a tumor - 724b23). All of this suggests that semen is a functional part (though not a bodily part in the way the uniform or non-uniform parts are - 724b29-32) of the organism and not a colliquation. In addition to this, Aristotle points out that there is a correlation between the amount of residue an animal secretes and the number of offspring it generates. Those that produce a lot of semen, even though they are small animals like fish, produce a lot of offspring. The converse is true with colliquation. Those animals that produce a lot of colliquation produce few offspring. That it is a residue of blood or final nourishment is evidenced by the fact that even after the secretion of a little semen, a feeling of great release occurs (725b6-9). Finally, there is the fact that semen does not occur in childhood or old age, nor in infirmity. The reason is that the organism does not (or barely) concoct a sufficient amount of nourishment in the first place, so that no leftovers occur (725b19-23). Aristotle, therefore, reaches the opposite conclusion of what the defenders of (P) hold. The semen is not a colliquation coming <u>from</u> the bodily parts but a residue of final nourishment to be distributed <u>to</u> all those parts and, since blood is that from which animals grow, this explains how a new animal like its parents can develop out of the semen (726b15ff).

2) Aristotle argues that the female does not secrete semen in the strict sense of the word. What she secretes is an analogue of the male semen but not semen itself. We find the following argument in a slightly different order at 726a4. (i) A creature is strong if it has a lot of heat and weak if it does not. (ii) Strong creatures produce little but thoroughly concocted residue, while weak ones produce less that is less thoroughly concocted. (iii) The female is a weak creature. Therefore, (iv) The bloodlike secretion called "menstrual blood" is an analogue of the male's semen. This is also clear from the facts that the menstrual blood begins and stops flowing at about the same time in life as does the semen of the male (727a 5-10). That the female is less hot and her residue less well concocted is shown by the facts that she is

pale and that her blood vessels are less prominent (a10ff). Now since no animal can have two seminal secretions and since the menstrual blood is the analogue of male semen, it follows that the female does not discharge semen (a27-29).

3) Aristotle argues in chapter 21 of Book One that the bulk of the male's semen plays no crucial role in the generative process. He begins by pointing out that lots of things such as artifacts come from things like a carpenter without that from which they come becoming part of the product. It is therefore possible that the bulk of the semen plays no crucial role (729b15ff). Furthermore, it is quite likely that this is so because in some cases of generation the female puts part of herself inside the male. This suggests that the heat and dunamis of the male is by itself enough to effect concoction of the female material (723b21-27 and 729b25-35).

But the greatest support comes from the phenomenon of wind eggs and oviparous fish:

> What happens in the case of birds and the groups of oviparous fish is the clearest sign that the semen does not come from all the parts of the body and that the male does not produce a certain part, such that it shall remain present within that which is generated, but produces an animal only through the dunamis in the semen. ...For if a hen bird happens to produce wind eggs, and if after that she is trodden, while the eggs have not yet changed from all being yellow to become white (i.e., are still growing), then fertile eggs instead of wind eggs come to be. And if she is trodden by another cock while the eggs are still yellow, the whole group of young take after the cock that trod her last. Therefore some (breeders) make use of this manner of copulation in order to produce thoroughbred birds, changing [cocks between] the first and the later covering. This suggests that the semen is not mixed up with (the offspring) and not present within it and that it does not come from the whole. For if it came from both (cocks), the off-

> spring would have these parts twice. But the semen of the male, through its dunamis, causes the matter and the nourishment in the female to take on a certain character (poian tina). And the (cock or semen) which is introduced later is able to do this through heating and concoction, for the egg receives nourishment until it has grown. The same happens with the generation of oviparous fish. For when the female has laid the eggs the male sprinkles milt. And all those that were not touched do not become fertile eggs, so that the male does not contribute the quantity (poson), but the quality (poian) (729b33-730a23).

The argument is an indirect one. Obviously, in the cases recorded here, the male does play an essential role in the generative process. It is equally clear that he does not provide any material part of the new animal. For if this were so, the young should have two such parts, one from each male that trod the hen. But they do not have two such parts and it is clear that the milt sprinkled on the eggs does not become part of what develops inside. Therefore, the male provides none of the material of the offspring. Since we know his role is indispensable, we must conclude that he provides the form of the offspring. Not the bulk, but the dunamis is the key to the male's contribution.

All these arguments are not simply part of a discussion of yet another bodily part not treated in the Parts of Animals, i.e., the semen. Rather, they are part of a carefully orchestrated attack on (P) and a defense of the form-matter hypothesis as the best alternative. Right after the wind eggs passage quoted above, we meet the main conclusion of Book I:

> From what has been said it is clear that for those animals which produce semen it does not come from the whole (body) and that the female does not contribute to the generation of offspring in the same way as does the male, but that the male supplies the principle of movement and the female the matter (730a24-28).

Throughout these discussions, Aristotle has juxtaposed (P) and the form-matter hypothesis. I think there can be no doubt that it is a genuine scientific hypothesis. Just how good a one it is depends, of course, a great deal on the kind of problems raised by (P) and on how well this hypothesis solves them. The next chapter deals with these problems.

FOOTNOTES

[1]For example, in De Anima I, 5, Aristotle criticizes the view that the soul consists of the four elements. Later, in II, 5, he claims he can solve the problems raised by this view by making use of the potential-actual distinction.

[2]Suzanne Mansion, "Le rôle de l'exposé et de la critique des philosophies antérieures chez Aristote" in Aristote et les Problèmes de Méthode (Louvain: Publication Universitaires, 1961), pp. 35-56.

[3]Pierre Aubenque focusses on this passage in his discussion of Aristotle's concept of aporia ("Sur la notion aristotelicienne d'aporie" in Aristote et les Problèmes de Méthode, p. 6-19). He points out that the notion of aporia has scholarly implications for Aristotle (p. 7). Often, as in Nicomachean Ethics VII and Posterior Analytics II (p. 10 and 11), Aristotle gives a statement of the problems and makes the thorough investigation of them an integral part of his research. That is, these statements of the problems blend in with the research at hand (p. 18). Aubenque also discusses other senses or uses of aporia.

[4]G. E. L. Owen, "Dialectic and Eristic in the Treatment of the Forms," in Aristotle on Dialectic. The Topics, pp. 104-108.

[5]In Book II: 733b24, 735a30, 736a23-25, 740b3, 741a4, a6, 743b33, 744a1, 745a19. In Book III: 752a4, 754b20, 755b1, 757a14, 759a8. In Book IV: 770b30, 771b15, 776a9.

[6]I,2,716a5; II,1, 731b18; 8, 747b30; IV,1, 763b24. From now on, I shall use the abbreviation G. of A. to refer to the Generation of Animals.

[7]The fourth kind of archē listed in the Metaphysics is: "that from which something first comes to be, which is not present within it, and that from which motion and change naturally begin, as in the case of the child which comes from the father and the mother..." (V, 1, 1013a8-10).

[8]In his introduction to the Loeb edition of the G. of A., Peck claims that eidos means soul (xii-xiv). To say that the male supplies the form is, according to Peck, to say that he contributes the (sensitive) soul. Now it is true that Aristotle believed that the male

supplied the sensitive soul and I discuss this claim in Chapter 4. But I do not think that in Book I eidos means, or should be interpreted as referring to, the soul of the offspring. There is a difference between an animal being of a certain species and having, let's say, arms, legs, two ears, and a head, and an animal's having a sensitive soul. The latter may well be a prerequisite for and a cause of the form but in that case, there still is the difference between the cause and the effect. Hence, I think that in Book I, the meaning of eidos lies closer to that of morphē or shape than to that of soul. All Aristotle tries to do in Book I is to establish that the male is the originator of the shape of the offspring. That the male does so by contributing the sensitive soul is a different question and is not brought up by Aristotle till Chapters 3 and 4 of Book II.

[9]Charles Darwin, The Variation of Animals and Plants under Domestication (New York: Orange Judd & Company, 1868), Vol. 2, Chapter XXXVII, "Provisional Hypothesis of Pangenesis." Darwin's defense of this hypothesis is very much like the one given by the defenders of (P), whom I shall identify shortly. Both argue that this hypothesis provides the best explanation of the phenomena to be accounted for, such as resemblances among members of species, etc. I have no idea whether Darwin, who did admire Aristotle greatly, hit upon this way of explaining heredity by reading the G. of A.

[10]My references to these treatises are in terms of the numbers assigned to the paragraphs by E. Littré in his Oevres Complète D'Hippocrate (Paris: J.B. Baillière, 1851), Vol. vii, pp. 470-485. The translations I use are those of Tage U.H. Ellinger's Hippocrates on Intercourse and Pregnancy (New York: Henry Shuman Inc., 1952). Ellinger followed Littré's text and retained his numbering system.

[11]Chapter 17 of Book I and the first half of 18 (where (P) is criticized) are, of course, full of references to (P). Others are: I, 724b35, 725a22, 776b10, 729a4, 729b33; II, 740b15 (?); IV, 1, 764b10, 766b10; 4, 769a14.

[12]Recently (Phronesis, Vol. XVIII, No. 1, 1973), Owen Kember has argued that Anaxagoras also held (P). He appeals to G. of A. 769a28 as part of his evidence. In that passage, Aristotle says that "there are some who say (tines hoi phasin) that the semen, although one

thing, is a certain panspermian of all (the parts)." Kember believes that we can include Anaxagoras among these unnamed physiologers. His main reason is that Alexander, in his commentary on a passage of De Sensu, notes that Anaxagoras believed that water contained juices, and, of course, Aristotle explains the panspermia theory as a mixture of many juices. Concludes Kember, "It can be reasoned therefore that Anaxagoras possibly associated panspermia and juices. From this it is a short step that the heredity theory, outlined at 769a28ff (with its references to panspermia and juices), properly belongs to Anaxagoras and perhaps some other (unknown) theorists" (pp. 11 and 12). In the absence of any fragment which directly attributes the pangenesis theory to Anaxagoras (and there is none), this cannot be a short step. In fact, it is a shot in the dark.

[13] Erna Lesky, Die Zeugungs - und Vererbungslehren der Antike und ihr Nachwirken (Wiesbaden: Franz Steiner Verlag GMBH, 1950), section D ("Die pangenesislehr) 1, "Ihr Ursprung aus dem atomistischen Denken." (Akademie der Wissenschaften und der Literatur, Abhandlungen der Geistes und Sozialwissenschaftlichen Klass, Jahrgang 1950 NR. 19).

[14] The two treatises occur on ancient lists of Hippocratic works but are not necessarily ascribed to Hippocrates himself. See Hippocrates (Cambridge: Harvard University Press, 1972), Vol. I, translated by W. H. S. Jones, p. XXXIX (Loeb Classical Library). Littré too has made these treatises part of the Hippocratic collection.

[15] A. L. Peck, tr. The Generation of Animals (Cambridge: Harvard University Press, 1963), note on page 50 in connection with 721b10 (Loeb series).

[16] Lesky, p. 72

[17] See Note 14.

[18] Gilbert Ryle, "Dialectic in Academy," in Aristotle on Dialectic. The Topics. ed. G. E. L. Owen (Oxford: Clarendon Press, 1968), p. 73.

[19] This list was suggested to me by the reading of Owen's "Dialectic and Eristic in the Treatment of the Forms," in Aristotle on Dialectic. The Topics (Oxford: Clarendon Press, 1968), p. 105.

[20] See Note 11.

[21] In the first sixteen chapters, he argues that the male and the female are the archai of genesis. From there on in to the end of the book he tries to show how each of the sexes is an archē and the specific manner in which each performs the task of being an archē. That is, he seeks to establish the dunamis and ousia of each. Aristotle seems to make the distinction between the that and the how at 716a7-18, where he tells us that a discussion of the nature and origin of semen will show the precise way in which the male and the female are the archai. This is, I believe, a forward reference to the discussion starting in Chapter 17. Before that, Aristotle has done nothing more than show that the distinction between the sexes and the possession of different procreative organs is almost universal. The claim that the male and the female are the archai of generation is a simple matter of observation. But the sense in which and how they are such must be established dialectically. See also II, 4, 738b20-22.

[22] 722a3-16, a17-b3, b31-723a9, b3-8, b16-18.

[23] 722b5-6, b8-30, 723a23-b2.

[24] 722b4-5, 723a9-23, b9-15, b19-27, b28-32.

[25] Charles Singer, A History of Biology (London and New York: Abelard-Schuman, 1959), p. 9.

[26] Joseph Needham, A History of Embryology (New York: Abelard-Schuman, 1959), p. 38.

[27] Needham, p. 39.

[28] Thomas East Lones, Aristotle's Researches in the Natural Sciences (London: West, Newman & Co., 1912), p. 195.

[29] Needham, p. 40.

[30] Sir David Ross, Aristotle (London: Methuen & Co., Ltd., 1966), p. 118.

[31] Marjorie Grene, A Portrait of Aristotle (Chicago: The University of Chicago Press, 1963), Chapter III, Section 6.

[32] Baruch Brody, "Towards an Aristotelian Theory of Scientific Explanation," in Philosophy of Science,

March 1972. Brody believes that Aristotle's essentialism can help block certain counter examples to the D-N model. A good explanation, he thinks, needs to meet all the D-N requirements plus the explanans must contain a statement describing (assigning) an essential property to a class of objects mentioned in the explanandum.

[33] This is taken from the title of the book Aristotle, Dean of Early Science by Glanville Downey (New York: Franklin Watt's Inc., 1962), p. 100. The author quotes George Henry Lewes with approval: "(The G. of A.) is an extraordinary work. No ancient work, and a few modern works, equal it in comprehensiveness of detail and profound speculative insight. We find there some of the obscurest problems of biology treated with a mastery which, when we consider the condition of science at that day, is truly astounding."

[34] Charles Singer, Greek Biology and Greek Medicine (Oxford: Clarendon Press, 1922), p. 31.

[35] Needham, p. 43.

[36] Ingmar Düring, "Aristotle's Method in Biology," in Aristote et les Problèmes de Méthode (Louvain: Publications Universitaires, 1961), p. 220.

[37] G. E. R. Lloyd, Aristotle: The Growth and Structure of His Thought (Cambridge University Press, 1968), p. 84.

[38] John Stuart Mill, Philosophy of Scientific Method (New York: Hafner Press, 1974), p. 265.

[39] Mill, p. 262.

[40] Karl Popper, Conjectures and Refutations (New York: Harper & Row, 1963), p. 53.

[41] Popper, p. 230.

[42] Mill, p. 261.

[43] This is D. M. Balme's translation, taken from Aristotle's De Partibus Animalium I and De Generatione Animalium I (Oxford: Clarendon Press, 1972). See also 724a7-11: "Further if the female does not emit seed, by the same argument it does not come from all. And if it does not come from all, it would not be inconsistent that it should not come from the female but that the female should be a cause of generation in some other

way." (Balme's translation)

[44] "There are a multitude of repetitions, and the order is haphazard, so that long digressions from the main argument are common." (Needham, p. 39)

[45] The spirit of eristic is contentious and the participants argue for the sake of arguing.(Topics 100 b23-101a4, 108a33-37, 112a9-11, 159a30-37, 161a23-24, 33b5. Sophistical Refutations 171b6-34; See Owen's "Dialectic and Eristic" p. 107, note 2.) The remarkable thing is that in Topics VIII, where Aristotle lays down the rules for participants in a serious dialectical dispute (161a37-61), he seems to allow for quite a bit of eristic argumentation. He advises that the questioner should leave the questioned party in the dark about what thesis he is trying to establish. Like Socrates, the questioner should seek certain seemingly unrelated admissions from his opponent and then at the very end he can bring the premises together and surprise his opponent with the conclusion (Topics VIII, 1, 156a9-13, a23-27, b31; 163b34). In the meantime, the questioner should use lots of devices to draw the attention away from his ultimate thesis so as to gain ready acceptance for the preliminary premises (Topics I, 18, 108a30-37; VIII, 1, 156b4, 157a1-4). Given this sort of advice, even a serious dialectical discussion can turn out to be quite a secretive affair. And it is possible to look at Aristotle's discussion of semen in just this way. Although the dispute commences in Chapter 17, we do not discover until Chapter 21 what Aristotle is driving at (730a24-28). It is only toward the end of the discussion that the major premises are brought out in the open, that the female does not secrete semen (I, 19, 727a26-30; 30, 728a26, a32; 729a21), that her menstrual blood is nourishment or matter (I, 19, 727b15; here Aristotle shifts imperceptibly from trophēn to hulēn, which, given the flow of the argument, is a crucial step. See Topics I, 18, 108a22-37), and that the bulk of the semen of the male plays no role in the generative process (I, 21, 729b19; 730a24-29; 21, 730b9-25). The conclusion has been hidden till the very end which is in keeping with the advice of Topics 151b31. Aristotle does not announce at the start that he wants to establish the form-matter hypothesis, but seems to prepare the way step by secretive step. First we are told that it is not unreasonable to suppose that the female does not secrete semen (721b13); then much later it is argued that she in fact does not secrete semen (727a26-30), and finally we learn that these admissions are stepping stones to a so far hidden conclusion that

the male supplies only the form and the female only the matter (729bff). In the meantime, we are kept busy with the various senses of the word "from" (I, 18, 724a21ff) and a lengthy discussion of why semen is a residue and not a colliquation (724b24ff), both of which are not in any direct way said to be relevant to the main thesis. We discover that later. The final conclusion comes very much as a surprise which is precisely what Aristotle recommends in the Topics (VIII, 1, 156a7-23).

This interpretation of the dispute is not faulty because it relies on Topics I and VIII indiscriminately. I do not think there is a great difference between the dialectic discussed in each book. Paul Moraux ("La joute dialectique d'apres le huitieme livre des Topiques" in Aristotle on Dialectic. The Topics, p. 130) holds the same opinion. He believes that Aristotle composed Books II-VII and Book VIII at different times and then added Book I as a general introduction when he put the two separately composed sections together.

After much prodding from Terry Penner, I gave up the interpretation I have outlined here because while it saves the integrity of the text of the G. of A., bringing the various elements together under a unified theme, it does not save Aristotle's reputation as a scientist. The interpretation I have given saves both Aristotle's reputation and the integrity of the text. I have already argued that the form-matter hypothesis is a genuine hypothesis offered by Aristotle as an improvement over the defective pangenesis theory. What may seem to be digressions of the main topic are actually various ways in which Aristotle defends his hypothesis. For the main topic is not a discussion of the origin and nature of semen, but the establishment of the archai of biology, the male as the supplier of the form and the female as the contributor of the matter. Most of what Aristotle says in Book One is directly related to this thesis. Below, I list three steps involved in the defense of the form-matter hypothesis. There are others such as the survey of the different meanings of "form" given at I, 18, 724bff.

CHAPTER III

FOR AND AGAINST PANGENESIS

The "for" and "against" of this chapter can be understood in more than one sense. On the one hand, these words have a straightforward connection with the text of the _G. of A._ In chapter seventeen of Book One, Aristotle tells us what evidence might be advanced on behalf of (P), while in chapter eighteen, he catalogues fifteen objections to pangenesis. In my two divisions of this chapter, I more or less follow the text.

On the other hand, the "for" and "against" also have some philosophical significance. In his list of objections to (P) (in chapter eighteen), Aristotle responds to the alleged evidence of chapter seventeen. I have pulled some of these responses forward to section A so that I discuss the evidence for (P) and the crux of Aristotle's response at the same time. Since there are a great many objections to (P) that are not closely tied to the evidence presented earlier, the objections do deserve separate treatment.

This chapter then is primarily a discussion of the weaknesses, defects, and possible strengths of the pangenesis theory. As such, it sets the stage for chapters IV and V where I discuss Aristotle's own solutions to the problems raised here.

Since my thesis that Aristotle argues with a dialectical opponent is in part a historical one, identifying this opponent with the author of _On Semen_ and _On the Nature of the Child_, there is no way I can avoid repeated quotes from these two treatises. It is the only way I can show that the link between Aristotle and our Hippocratic author is not a mere coincidence, but that Aristotle seems to have written large parts of the _G. of A._ with the arguments from these two treatises in mind.

A. For Pangenesis

Right after Aristotle introduces the pangenesis theory in chapter seventeen of Book One, he says this:

> There are about four good pieces of evidence (_tekmeriois_)[1] someone could use to prove that the semen is drawn from each of the parts. (721b13-15)

In this section, I discuss these pieces of evidence and Aristotle's response to them. In three of the four cases, I also establish the link with On Semen. I have labeled the arguments as follows: 1) The argument from pleasure; 2) The argument from mutilations; 3) The argument from resemblance; and 4) The argument from the whole to the parts.

1. The argument from pleasure (271b15-17)

"First there is the intensity of the pleasure. For the same affection becomes more pleasant when there is more of it and that which occurs in all the parts is more than what occurs in either one or a few." Formalized, the argument reads like this: 1) the intensity of a certain experience increases as the scope or number of bodily parts affected increases; 2) The intensity experienced in copulation is enormous. Therefore, 3) the semen, the ejection of which is accompanied by this intensely pleasant experience, must be drawn from all the parts of the body.

Aristotle has accurately reported the views of our Hippocratic author. For in §1 of On Semen, in the context of explaining pangenesis, the author gives the following proof:

> Here is an illustration in proof that it [the semen] is a separation of the most potent part. The discharge of such a small quantity makes us weak. This is the way of it. Veins and cords lead from every part of the body into the penis; when these are somewhat affected by friction and are heated and filled, a tickling sensation (*knesmos*) sets in and consequently, pleasure and warmth pervade the whole body. As the penis is rubbed and the whole person is agitated, the fluid in the body is heated, diffused and churned by the movement and becomes foamy by churning.

Actually there are two arguments here. One seeks to explain the great weakness experienced after the ejection of semen as being due to one's loosing a most potent protion of each bodily part. And the other argument explains the pleasure which permeates the whole body as being due to the same phenomenon. It is interesting to note that the Hippocratic author--though he did believe that the female too secreted portions from all her bodily parts--did not offer the exact same argument from pleasure in the case of the woman. He thought the man's pleasure was more intense and then adds "as regards the woman in copulation I assert that when the genitals are rubbed and the womb agitated, there occurs in it a sort of tickling sensation (knesmos) and the rest of the body derives pleasure and warmth from it" (§ 4).

It is almost uncanny how Aristotle picks up every one of the points made by the Hippocratic author. In the middle of his own defense of semen as a residue of final nourishment, he mentions the paradox "that although only a little quantity of semen is emitted the exhaustion is conspicuous" (725b7). He takes this as proof of the fact that semen is indeed what he has argued it was, namely, a residue of final nourishment which was about to be distributed to all the bodily parts. The exhaustion is due to privation of nourishment.

Aristotle responds to the pleasure argument by offering a different cause for the intensity of the experience.

> The cause of the intense pleasure that occurs when two lovers engage in intercourse, is not that it comes from the whole, but there is a strong itching (knesmos ischuros).[2]...Further the pleasure comes at the end, while it should occur in each of the parts, and not at the same time, but in some earlier and in others later. (723b33-724a4)

The strong itching, rather than the fact of (P), is suggested as the cause of the intensity. Later on in the discussion, he turns the Hippocratic claim that the pleasure experienced by the woman is not quite like that of the man's to his own advantage. At 727b7, he argues that the female sometimes experiences no pleasure at all while presumably the male always does. He takes this as support for his own belief that the female does not secrete the same kind of semen as does the male. And at 728a34, he argues that the woman does not secrete semen because she experiences pleasure at the place the male touches her, but that is not the place where the fluids are emitted. Thus Aristotle has taken the argument from pleasure, which the Hippocratic author offered as a piece of evidence for (P) and turned it into evidence for the form-matter hypothesis.

2. <u>The argument from mutilations</u> (721b18-20)

"Further, mutiliated offspring comes to be from mutilated parents. For they say that when the parent is deficient in a part, no semen comes from that part and its correlate in the offspring does not get formed." This line of argumentation resembles very closely the eleventh paragraph of <u>On Semen</u>. I quote the last half of it:

> But if the parent suffers from disease, and if the four congenital fluids which supply the semen, but semen with that portion of it weaker which comes from the crippled part, then it does not seem astonishing to me that the child should be crippled like the parent.

Aristotle has two criticisms of this argument (724a4-7). One is that it is not true that mutilated parents always produce mutilated offspring. This is a curious objection in light of the fact that the Hippocratic author openly admits this in the first line of the paragraph from which I just quoted. It reads: "With regard to children of crippled individuals, they are born normal in most cases; for the crippled part has its full quota when the parent is healthy." Aristotle seems to have conveniently ignored this first sentence. His second point of criticism is that the cause of mutilation is the same as the cause of resemblance. Since (P) cannot be the cause of the latter, it also cannot be the cause of the former. This brings us to the third and the most important piece of evidence for the fact that (P).

3. <u>The argument from resemblance</u> (221b21-24)

This is by far the most potent piece of evidence, and Aristotle spends a great deal of time trying to refute it. I discuss that argument at some length in Chapter V. Here I shall only concentrate on the main points. This is the argument:

> Besides these [mutilations] there are [positive] resemblances with respect to the parents. For offspring comes to be which resembles [the parents] not just with respect to the whole body, but also part for part. If then the cause of the resemblance of the whole is that the semen comes from

> the whole, then the cause [of the resemblance] of the parts would also be that something comes from each of the parts. (721b21-24)

The Hippocratic author is presented as arguing that since the resemblance between parents and offspring as wholes is due to the semen as the only link between generations, that, therefore, resemblance in part must be due to that part of the semen which was drawn from the appropriate parental part.

What the Hippocratic author says in §8 can easily yield this kind of argument. The first lines, for instance, clearly link (P) to the resemblance between parts:

> And in the semen itself there is some drawn from each part of the body, both of the woman's and the man's; and weak semen from weak parts, and strong from strong parts. There is necessarily a corresponding result in the child. From whatever part of the man's body more is contributed to the semen than from the woman's, in the offspring that part is a better likeness of the father; from whatever part more is contributed from the woman, that part in the offspring is a better likeness of the mother.

The last line raises the obvious question of how Aristotle himself, given his form-matter hypothesis, can handle these kinds of resemblances between the offspring and the mother. I discuss this in chapter V.

Aside from this problem of how he explains these phenomena, Aristotle has an impressive array of arguments against this piece of evidence. Of the fifteen objections he brings against pangenesis in chapter eighteen, by far the largest single group centers on this issue. A very obvious line of attack is the raising of counter examples. Aristotle simply argues that not all cases of resemblance can possibly be due to the fact that the semen came from the parts involved in the resemblance.

> Because children come to be [people] who resemble their parents in voice,

> hair, and nails and even in the way they move, but nothing is drawn from these parts. And the parents do not yet have certain characteristics when they generate, for example, gray hair or a beard. Further, children resemble their remote ancestors, from whom nothing is drawn. (722a5-8)

Within one stroke, Aristotle has undercut the argument from resemblance. All three points seem to be good ones. There are resemblances between parts or aspects of a parent and a child from which obviously no semen came. The Hippocratic author does say that portions of the semen come from <u>all</u> the parts of the body. But this still leaves unexplained resemblances in bodily movements, those concerning characteristics the parents did not have at the time of generation, as well as those with remote ancestors. As we shall see later, in order to account for these last sort of resemblances, Aristotle does quite a bit of theoretical maneuvering.

4. <u>The argument from the whole to the parts (721b24-28)</u>

> Further it also seems reasonable to suppose that just as there is something in the case of the whole from which it first came to be, so there is something in the case of each of the parts. And since that [first] something is a semen, the something related to each of the parts could also be a special semen.

There is no analogue for this argument in <u>On Semen</u> and to my knowledge Aristotle does not specifically respond to this argument either.[3] But this lack of an exact correlation does not obstruct my general thesis.

I suggested in the preceding chapter that perhaps this discussion in the first book of the <u>G. of A.</u> was an echo of an ongoing debate in the Academy about the <u>archai</u> of biology. For that to be the case, there need <u>not be</u> an exact correlation between Aristotle's report of how (P) can be defended an the way it is actually defended in <u>On Semen</u>. We can safely conclude that the theories expounded in <u>On Semen</u> provided the raw material for pieces of evidence recorded in this chapter of the <u>G. of A.</u> As usual, Aristotle elaborated and systematized what he found in his predecessors or contemporaries.

These pieces of evidence also show that Aristotle did not set up a straw man. The dialectical dispute is a genuine one with good arguments on both sides. Aristotle may have been successful in his refutations of the arguments from pleasure and from mutilation but the one from resemblance is not so easily overcome. The fact that children resemble their mother as well as their father is both strong evidence for something like (P) and the single most effective objection to the form-matter hypothesis.

B. Against Pangenesis

A dialectical discussion is supposed to commence by the raising of difficulties connected with the endoxon that has been selected as the point of departure (Topics I, 2, 101a34-36). Aristotle does precisely this when, in the eighteenth chapter, he lists his objections to the pangenesis theory. Also, a questioner must seek to establish the opposite of what the answerer believes (Topics VIII, 5, 159b6). By maintaining that the semen is a residue of final nourishment about to be distributed to all the parts, Aristotle does indeed establish the exact opposite of what the defenders of (P) hold. They believed that it came from all the parts (cf., G. of A., I, 18, 722a2). But to establish the opposite, Aristotle must destroy the plausibility of (P) and these objections are designed to do just that. It is also the task of "the questioner to so conduct his discussion that he makes the answerer say the most unlikely things, which follow necessarily from the thesis" (Topics, VIII, 4, 159a18-20). Before his critique, Aristotle announces that "the arguments presented are not difficult to refute and, besides that, the theory has impossible implications" (I, 18, 722a3).

All in all, there are fifteen separate objections, each introduced with eti.[4] Some of these have already been discussed in section A. I have grouped the remainder under three headings: 1) The problem of resemblance; 2) The problem of development; and 3) The problem of sex differentiation. Though I discuss these arguments with an eye on Aristotle's own solutions to be discussed later, these arguments do, as Ross[5] has pointed out, have an intrinsic merit. They show Aristotle's critical abilities at their best.

1. The problem of resemblance

In the preceding section, I already mentioned some of the counter-examples to (P). These were cases in which there was a resemblance, but no semen was said to be drawn from the parts involved. In his eighth objection, Aristotle gives a counter-example in the opposite direction. He suggests that there are cases of generation that involve semen, but where there is no resemblance between offspring and parents. Flies and fleas, for instance, do not produce animals of their own kind, but larvae (723b4). Therefore,

> It is clear then that all those that are of a different kind [than their parents] do not come to be because it was drawn from every part. For, if resemblance is a sign that it is drawn from the whole, then they would have to be similar [to their parents]. (723b6-8)

The defenders of (P) might deny that there is semen involved here. Aristotle would then have to show the opposite to be the case. But that would not be easy, for he has said at 721a13 that "the males [of flies and fleas] do not seem to have seminal passages." Further, Aristotle's suggestion that semen is involved in these cases conflicts with his definition of semen as concocted blood, these insects being grouped among the bloodless animals (I, 1, 715b, and III, 9). Clearly, the counter-examples mentioned earlier are more effective than this one.

A major objection to the argument from resemblance is based on Aristotle's own schema of the biological world (722a17-722b3).[6] The defenders of (P) must claim that the semen is drawn from all the uniform parts only, or from all the non-uniform parts only, or from both. In all three cases, they run into trouble. If the semen is said to be drawn from the uniform parts only, then they are faced with the fact that the resemblance is often far more striking in the non-uniform parts, like a hand or a face. Since these latter cases of resemblance must be due to a cause other than (P), "why does it not follow that the other cases of [uniform] resemblance are also not due to the semen's being drawn from the whole, but to some other cause?" (a22)

If the defenders retract and claim that the semen comes only from the non-uniform parts, a similar problem arises. Whatever is the cause of resemblance on the uniform level is, the same might well hold on the other level. Further, this suggestion is strange because the non-uniform parts are made up of the uniform ones. So how can the semen be drawn from the one and not from the other?

It would seem then that the semen is drawn from both sorts of parts. This too is problematic because each level of parts (uniform, non-uniform, the whole) is made up of the preceding level (the four elements, the uniform parts, etc.) plus a synthesis (sunthesiōs - a29) or composition. Since this synthesis or composition is not itself a part, but the organization of the parts, (P) cannot explain resemblance because the semen is not drawn from the major cause of the resemblance, namely, the synthesis. In a passage reminiscent of Metaphysics Z, 17, he explains it as follows:

> It is like when something is drawn from a written word, if it comes from the whole word then it comes from each of the syllables, and if from these then also from the letters (stoicheiōn) and their synthesis. (722a31-33)

Understanding is reached not of the components taken singly and in isolation but as a result of the synthesis and juxtaposition of the parts. And so it is, says Aristotle, with the origin of resemblance. Whatever it is that creates the synthesis--which is the real cause of resemblance--that, and not (P), is the cause of resemblance (a35-b3). The fact that the semen is drawn from all the parts cannot explain resemblance because it is not drawn from the synthesis, which is not a part and yet the real cause of resemblance.

An interesting objection to the argument from resemblance (and the final one to be discussed) is based on the phenomenon of transplants (723b16-18). At the very beginning of the G. of A., Aristotle announces that plants will have to be dealt with separately (716a1). Professor Balme believes that this remark is "probably inserted by an editor."[7] The reason he gives is that we don't know whether Aristotle wrote a treatise on plants so that references like these more likely than not were inserted to refer to Theophrastus' botanical works. I suggest, pace Balme, that this remark and other references[8] to plants fit very well in the scheme of the G. of A., and they do so without supposing that Aristotle wrote a treatise on plants.

The reference at 716a2 does not say that Aristotle wrote a treatise on plants, but that plants will have to be dealt with separately. This second statement of disavowal is a very natural one once we realize that in _On the Nature of the Child_ (§22-§26), the Hippocratic author had extensively discussed the conditions under which plants grow with an eye to explaining the generation and development of animals. But Aristotle, though he takes _On Semen_ and _On the Nature of the Child_ as his model, is not going to treat both plants and animals in one treatise. He made the remark at 716a2 by way of apology for deviating from his model. I think this is a better explanation than one that relies on a mysterious _ad hoc_ editor.

Aristotle's reference to transplants, at 723b17, supports the explanation I have given. In §26 of _On the Nature of the Child_, the Hippocratic author says this about transplants:

> As for buds taken from one tree and inserted in another, which, having become trees on trees, live and bear fruit different from that of the tree in which they are inserted, this is how it happens.

The point is that the transplanted part, which at first receives nourishment from the tree on which it has been grafted and later from the ground, nevertheless bears its own kind of fruit instead of the kind that grows on the tree onto which it was grafted. Aristotle seems to refer to this difference when he says that these "transplants bear a seed from themselves" (_ap'auton_ - b16)

He uses this phenomenon of transplants to argue against (P):

> It is clear that even before they were transplanted they bore fruit from the same quantity [of the plant] and that the seed does not come from the whole plant. (723b17-18)

The fact that the transplanted part bears its own kind of fruit after it has been transplanted means that its seed is not drawn from the whole of the tree on which it was grafted (nor, of course, from any part of it). By the same token, its seed need not have been drawn from the whole plant before it was transplanted. If the transplanted part can bear fruit of its own after the cutting, then it can also do so before and its seed need not have been drawn from the whole. Like in the argument from pleasure, Aristotle has taken part of the evidence advanced by our Hippocratic author and turned it against him.

2. The problem of development

One of the main problems with the pangenesis theory has to do with the obvious, living unity that an animal is. As Aristotle sees it, there is a gap between the input into the semen (portions from all the bodily parts) and the output (a living, unified animal). Since all it deals with is the input, (P) cannot be used to explain this difference. Asks Aristotle, "If the parts exist in the semen in a disconnected state, how can they be alive? And if they are connected there would have to be a small animal" (722b4). The latter option conflicts with the observed facts (cf., G. of A. II, 740a17, 741b25; History of Animals VI, 3, and 10) and the former are not really parts of animals, but only homonymously so (722b17, Prior Analytics I, 640b29). Aristotle himself has difficulties with this problem. For even if the form-matter hypothesis is correct, that still leaves unexplained how the form gets to be gradually imposed on the material. He handles this with the aid of his potential-actual distinction. I discuss this in the next chapter. For now I want to stress the fact that Aristotle regards the inability to explain the origin of a living and unified animal as a very serious defect of (P).

He also has a more or less internal objection to (P), involving some of the views Anaxagoras held. Anaxagoras believed that flesh could come from bone and bone from flesh, that each one of the uniform parts contains portions of all the others so that one of them could be the source of all the others. Similarly, the defenders of (P), says Aristotle, should argue that all the bodily parts (somehow) come from one bodily part, the semen.

In fact, their case of the semen's coming from all the bodily parts is lost anyway, so they might as well adopt an Anaxagorian kind of position. It is lost because all the bodily parts receive their names from either functions (as in the case of the non-uniform parts) or from their texture, their softness or hardness (as in the case of the uniform parts). Now, "it is clearly impossible that that which is drawn from all the parts should have the same name as those parts" (722b35). The semen is not called "flesh," or "blood," or "nose," or "hand," because it does not have the characteristics of those parts. It is not called "blood" because it is not blood. (This conflicts, of course, with Aristotle's own redefinition of semen as a residue of blood, but not with his discussion of the nature of semen in II, 3). Or, to put it the other way around, the semen is not blood because we don't call it "blood." Still, the offspring with all its parts, including its blood, comes from the semen. This means that blood comes to be from what is not blood.

> But if blood [of the offspring] comes to be from something which is different, then the cause of resemblance would not be, as the supporters suggest, that the semen is drawn from all the parts. For it is sufficient [to explain resemblance] that it come from one part only, if blood does not come from blood. For why should not all the parts come to be from that one thing? (723a2-6)

And what goes for blood can go for all the other parts. So why not have the semen as the one part from which all the others somehow come? There still is a difference between Anaxagoras' view and pangenesis, for whereas he held this sort of belief with respect to all things, the defenders of (P) need only hold it with respect to the generation of animals (738a5).

But even this difference has to be shrunk once the defenders of (P) face up to the "somehow" that still lingers in the background. It is because they could not see how else a fully grown animal could come from the semen, that they supposed the semen to contain all the parts of that animal. But how does an animal get to be fully grown? Again, Anaxagoras makes sense:

> Further, in what way shall these [parts] that are drawn from the whole grow? For Anaxagoras makes sense when he says that the flesh from the nourishment is added to the flesh (723a9-11).

Anaxagoras could account for the phenomenon of growth because he believed that the nourishment an animal takes in has within itself portions of all the uniform parts, such as blood, flesh, and bone. The defenders of (P) have to either take a position like Anaxagoras, or they have to admit that what is taken in as food is (somehow) changed into the bodily parts. Aristotle takes them to be believers in change (723a8) and closes in on them with this question:

> But if it is possible that that [food] which is added changes, why can it not be the case that right from the start the semen is such a thing, that from it blood and flesh can come to be, even if it itself is not blood and flesh? (723a14-17)

The idea is that there is no difference between allowing for a substantial change of food into bodily parts and for allowing the semen to change into such parts; the changes are equally drastic. Since the defenders of (P) allow for the first change, they might as well allow for the second. And they will not be able to claim that the miniature parts of the semen grow by mixing with food "the way wine increases in bulk when water is added" (a19). This is impossible, for each of the bodily parts would have to be present in the semen before the mixture took place. This, however, conflicts with the observation that many of the parts--such as flesh and bone--develop later. Furthermore, "The claim that a certain part of the semen is sinew and bone is far beyond us" (a22), says Aristotle. Growth simply cannot be explained in terms of mixture.[9] Consequently, the point of having to admit a drastic substantial change from the semen to the bodily parts that develop out of it still stands.

Note that Aristotle has not said that Anaxagoras held the pangenesis theory. On the contrary, he has used Anaxagoras' view to attack and improve upon that theory. The correction is twofold. First, the defenders of (P) could avoid some of the obvious counter-examples if they claimed that the semen contained all the parts without claiming that it came from all the parts. One bodily part, the semen, could contain all the others, the way Anaxagoras said one of the uniform parts contained all the others. Second, having gone this far with Anaxagoras, they should either go all the way and admit that food too, just like semen, contains all the parts of the body, or they should allow for a radical and substantial change. Since they obviously think the food changes, they should admit that other, just as drastic a change, also. Thus, they are forced into accepting Aristotle's position that the generation of an animal from the semen involves a drastic change and the development of new structures as over against the mere unfolding (in the sense of growing bigger) of already present parts.

3. The problem of sex differentiation

The Hippocratic author believed that "a mixture is affected of that which comes from the man and that which comes from the woman"(On Semen, §5). He had quite a bit to say about how this mixture was affected (see chapter IV), but Aristotle (conveniently) ignores these complications and interprets the idea literally. He asks: "and what about the generative organs? For that which comes from the male and that which comes from the female are different" (722b5). Taken literally, the pangenesis theory leads to there coming to be two animals out of each mixture, a male and a female or, if just one, it will have to have both male and female reproductive organs. Further, the most obvious question, given (P), is why there need to be two sexes. If the semen of the male comes from all the parts of his body, why does he need the female to reproduce? Similarly, why does the female need the male? (722b8)

Again, the defenders of (P) could learn something from a pre-Socratic philosopher. For Empedocles too thought generation was due to a mixture, but he believed that "in the male and the female there is something like a "sumbolon" and that the whole is not drawn from either of the parents" (722b11-12). A sumbolon refers to a half of something that is used. Liddell and Scott give the example of two pieces "of a coin which two contracting parties broke between them and each preserved one part." So, Empedocles believed that the male and the female each had complementary parts.

This solution is impossible (722b17) because separated parts of an animal, whether big (as in the case of Empedocles) or small (as in the case of pangenesis), cannot be alive.

> For either, if [these separated parts] do not have a soul or life or some such power, they cannot remain alive, or if they already are more of an animal, they cannot grow into becoming one animal. (b21-24)

If the male and female each have half of an offspring, how can those two halves live before they are joined?

Empedocles' view also provides a possible answer to the question of how the offspring gets its sex. He supposed that the male and the female are differentiated during conception (723a24). Aristotle thinks this is quite plausible because it is a known fact that females change from being fertile to being infertile and from producing males to producing females (a27), all of which suggests that a certain female's semen does not contain a miniature female reproductive organ. It all seems to depend upon the circumstances of conception.

> It is clear that, if we suppose this to be so, namely, that the semen itself can become female and male because there is no sexual part present within the semen, that a female does not come to be on account of being drawn from a certain part and that there is not in the semen a part which is peculiarly male or female. What then is the difference between saying it about this one part and saying it about the others? For if no semen comes from the uterus, the same argument holds for the other parts as well. (723a32-b2)

Since a female's semen does not contain a portion from the uterus, it makes no sense for the defenders of (P) to insist that the semen does have portions of all those other bodily parts. Curiously enough, the author of <u>On Semen</u> records the exact same facts as Aristotle here reports, and uses them to conclude that the portions of male and female parts differ in strength, to which difference the differentiation of the sexes is ascribed(<u>On Semen</u>, §7). Aristotle ignores this last aspect of variation in strength and sticks to his literal interpretation of (P). Given that interpretation, his objections are good ones.

A problem related to that of the differentiation of the sexes has to do with multiple births. Aristotle wonders how the defenders of (P) can explain the fact that sometimes from one act of coitus more than one offspring comes to be.

> For it is necessary that one secretion comes from one act of coitus and from one act of secreting. It is impossible

> that it is divided up in the uterus, for in that case the division would already be as from a new plant or animal, not from the semen. (723b13-15)

The defenders of (P) have to hold that in the case of multiple births, the semen contains more than one portion from each of the bodily parts, which is unlikely, since we are talking about only one secretion. Or they would have to hold that once the semen arrives in the uterus, it divides itself up into several portions. This is indeed the explanation offered by the Hippocratic author. In §20 of _On the Nature of the Child_, he explains the birth of twins as follows:

> Twins are born from a single act of intercourse in this way. The womb has numerous and curved cavities, some further from, other closer to the genitals. Animals which bear many young have more cavities than those that bear few. It is so in sheep, wild beasts, and birds. When the semen happens to be divided by arriving in two cavities, and the womb has received it and neither cavity lets the semen flow into the other,then the semen surrounds itself with a membrane in each cavity and gains life separately in the same way as has been stated in the case of a single embryo.

Aristotle's objection is that in the case here described, "the division would already be as from a new plant or animal, not from the semen" (723b14). The crucial premise for the defenders of (P) is that an animal can come to be from the semen because the semen contains portions from all the parts. If it supposed that the semen divides itself up into the different cavities, in each of which a complete animal grows, then each of these divisions has to be looked upon as a separate act of coitus. We can't look upon it as a simple division of parts for that supposes that the semen has more than one portion of each bodily part and that more than one act of coitus lies behind the birth of twins, for each act of coitus is supposed to be matched by one act of secretion. Hence, in order to explain how one act of coitus can yield twins, each separation of the semen in the uterus has to be viewed as a separate act of coitus. This, of course, has the absurd consequence that the initial act of coitus (the real one) is followed by other acts of coitus in the uterus. But how can there be an act of coitus without there being also a new plant or animal in the uterus? All of this is absurd, and the birth of twins from one act of coitus is left unexplained.

These, as Aristotle sees it, are the main problems connected with the pangenesis theory: the problem of resemblance, the problem of development, and the problem of sex differentiation. In the next two chapters, I discuss Aristotle's own solution to these problems.

FOOTNOTES

[1]The word tekmerion is a technical word for Aristotle. He discusses the kind of argument to which it refers in Prior Analytics, II, chapter 27, and in the Rhetoric, I, chapter 2. Le Blond (p. 241, note 1) may be right when he claims that Aristotle often uses tekmerion and semeion interchangeably. But this does not mean that the technical meaning discussed in the Prior Analytics and the Rhetoric is never on Aristotle's mind. I think a good case could be made that here in the G. of A., I, 17, the word has the technical meaning of "necessary sign." Basically, a tekmerion is an argument in which one can validly infer a cause from its effect. For example, from the fact that 1) all those who have fever are sick, and the fact that 2) he has a fever, one can infer that 3) he is sick. The first three arguments I discuss do indeed have a structure like this. (Note that he says that there are about four--i.e., three--such arguments.) The idea is that that argument cannot be faulted logically. If the premises are true, the the conclusion necessarily follows. A tekmerion is not simply a "piece of evidence," as Balme has it in his Aristotle's De Partibus Animalium I and De Generatione Animalium I (Oxford: Clarendon Press, 1972), but it is a good piece of evidence. Balme suggests in his commentary (p.141) that there is a discussion of these four pieces of evidence in Lesky's Die Zeugungs--und Vererbungslehren der Antike und ihr Nachwirken (Wiesbaden, 1950), 129f. I have been unable to find a discussion of these four pieces of evidence in this work. Lesky does, of course, trace (P) to Democritus' atomism and sees that as the source for the theory in the Hippocratic treatises we have been discussing. But she does not discuss these pieces of evidence as evidence for (P), nor does she specifically link Aristotle's account in G. of A. I, 17 to what I have pointed out in On Semen.

[2]Note that the Hippocratic author uses the same word knesmos. See preceding quote in the text.

[3]Balme thinks Aristotle responds to this argument at 722a4-b3. I view the arguments presented there mostly as a response to the argument of resemblance. See the problem of resemblance in the next section.

[4]1 (722a3-16); 2 (722a17-722b3); 3 (722b4-6); 4(722b8-30); 5 (722b31-723a9); 6 (723a9-23); 7 (723a23-723b2); 8 (723b3-8); 9 (723b9-15);

10 (723b16-18); 11 (723b19-b27); 12 (723b28-33); 13 (723b34-724a3); 14 (724a4-7); 15 (724a8-13).

[5]Sir David Ross, Aristotle (London: Methuen & Co., Ltd. 1966), p. 119. See also G. E. R. Lloyd, Aristotle: The Growth and Structure of His Thought (Cambridge: The University Press, 1968), pp. 82-84.

[6]See Parts of Animals, II, 1.

[7]Balme, p. 131.

[8]E.g., II, 1, 733b24.

[9]See Generation and Corruption I, 5.

CHAPTER IV

THE POTENTIAL-ACTUAL DISTINCTION

In the preceding chapter, I discussed Aristotle's criticisms of the pangenesis theory. Even if we suppose these criticisms to be correct, this by itself does not establish the form-matter hypothesis. Aristotle must further show that the problems connected with (P) can be solved if we adopt his hypothesis. In this and the next chapter, I show how Aristotle does this. I begin with the set of problems connected with the question of how an animal develops from the embryo. In the next chapter, I tackle the problem of resemblance between parents and offspring.

Aristotle claimed that the adoption of (P) does not help us explain how a living, unified, and besouled animal develops from the embryo. We must note, though, that the form-matter hypothesis by itself does not explain this fact either. Clearly, the male is not in continuous contact with the female, the way a sculptor is always in touch with the marble and the carpenter with the wood. It is not at all easy to see how the male succeeds in imposing his form upon the female material. It is one thing to say _that_ the male does this as Aristotle does in Book I, but quite another _how_ he does it. Aristotle tackles this second problem in Book II. In so doing, he follows the division laid down by our Hippocratic author, who in _On Semen_ discussed the nature of semen, and in _On the Nature of the Child_ how a child develops from the mixture of two semen. Aristotle too first discusses the nature of semen (I, 17-21) and then the problem development (II, 1-3).

To answer the problem of development Aristotle makes use of his potential-actual distinction and it is on his introduction and use of that distinction that the discussion of this chapter focusses.

A. _The Distinction Introduced_

A fruitful way of looking at the origin of the potential-actual distinction is to view the rise of this distinction as a scientific revolution precipitated by Aristotle. I believe that the distinction owes its high ontological status in the central books of the _Metaphysics_ to its usefulness in solving problems in the philosophy of nature. The fundamental principle of explanation for most natural philosophers before

Aristotle was the maxim that like naturally goes to like. I shall call this maxim (L). As Aristotle saw it, (L) created at least three important problems: 1) the problem of how things like flesh and bone come to be out of the four elements; 2) the problem of how it is that we sense external objects; and 3) the problem of development. In each case, Aristotle more or less consciously replaces an appeal to (L) with an explanation based on his own potential-actual distinction. The first problem is solved in this way in Book II of _Generation and Corruption_, the second in _De Anima_ II, 5, and the third will be discussed in this chapter. My reason for mentioning these other problems is to draw attention to the fact that the shift in explanatory principles in biology is part of a larger revolution in natural philosophy.

The first book of _Generation and Corruption_ gives ample support to my thesis. In it, Aristotle discusses all sorts of change, generation, and destruction. At important points in his discussions, he points out that his predecessors used the like-goes-to-like maxim to explain phenomena such as growth (315b), coming to be and passing away (315b18), action and passion (323b7), and mixture (328a4). Aristotle (correctly) believed that those of his predecessors who had bothered to explain natural phenomena made indespensable use of the like-like maxim. I shall refer to this maxim simply as (L). If we, in contrast, look at Aristotle's own explanations of the same phenomena, we find him repeatedly using the potential-actual distinction (317b15, 320a11, 320b18, 326b30, 327a29). It is true that in this book, Aristotle never in the same passage draws an explicit contrast between the two models or paradigms. But the procedure he follows leaves no doubt that, in fact, explanations based upon the like-like maxim are replaced by ones based upon the potential-actual distinction.

I have called Aristotle's replacement of (L), with his own distinction, the external (philosophical) motivation for his introduction of this new distinction. After I discuss this external motivation, I also discuss an internal reason for the replacement. That is to say, even without the problems raised by an appeal to (L), there are good reasons within Aristotle's own philosophical perspective for incorporating the potential-actual distinction in an explanation of the development of animals.

1. The external motivation

My point here is both historical and philosophical. I show that our Hippocratic author used maxim (L) to explain the development of animals and that Aristotle consciously replaced that particular use of (L) with his own potential-actual distinction. Furthermore, he felt that the problems created by an appeal to (L) were indeed grave enough to warrant the use of his distinction. Though both of these points, one more historical and the other more philosophical, are made, I make no conscious effort to keep them separate.

I start with a long quote from On the Nature of the Child and want to draw special attention to the author's appeal to (L). The present passage contains the clearest such appeal in ancient embryology. My thesis is that Aristotle consciously replaces this appeal to (L) with an explanation based on his own potential-actual distinction. Here is the passage:

> As the flesh grows, it is differentiated by the air, and each kind of substance in it goes to its own kind, the solid to the solid, the spongy to the spongy, the fluid to the fluid; and each thing goes to a place of its own according to the kind from which it originally came. What came from the solid is solid; what came from the fluid is fluid; and the other kinds are produced in the same way during growth. The bones become hard, solidified by heat. Moreover, they also ramify like a tree. Both the inner and outer parts of the body also become better differentiated. The head becomes detached from the shoulders; the arms and the forearms from the sides. The legs separated from one another. Cords spread rapidly about the growth of the joints and are self-hardened. The nose and ears are detached amongst the flesh and become perforated. The eyes are filled with pure fluid. And the sex of the genitals becomes plain. The viscera become differentiated. Furthermore, there is respiration through both the mouth

> and the nose. The abdomen is filled with air, and the intestine, inflated from above by the air, cut off the breathing through the navel and stop it. A passage out is formed from the abdomen and the intestines to the anus; and a passage out into the bladder. Each of these parts is differentiated because of the respiration. All things separate by kinds when they have air blown into them. In fact, if you will attach a pipe to a bladder and introduce through the pipe into the bladder some earth, some sand, and some fine lead shavings, pour in water, and blow through the pipe, all these substances will at first be mixed up with the water, and then after a certain time, as air is blown in, lead will go to lead, and sand to sand, and earth to earth; and if one lets them dry and then tears the bladder open and looks, one will find that in each case like has gone to like. In the same way the semen and the flesh are differentiated, and each kind that is in it goes like to like. (§6)

The Hippocratic author believed that from the mixture of the two semen, flesh is the first thing that develops; it is simply a coagulation of the blood which is drawn in together with cold air. Out of this coagulated blood, the various bodily parts are differentiated by the process of breathing. The air is drawn in through the navel like when air is drawn into a bladder through a pipe. The material in the bladder, such as earth and sand, will separate itself into various sorts and places, earth going to earth, and so on. Just so, says the author, the content of the embryo is sorted out by the process of breathing. The solid goes to the solid, the fluid to the fluid, and "each thing goes to a place of its own, according to the kind from which it originally came."

The reference to fluids and solids raises a problem of interpretation. In the preceding chapter, we saw that Aristotle interprets the pangenesis theory quite literally as the belief that the semen contains portions from all the parts of an animal, uniform as well as nonuniform. From the above quote, it would seem, however, that the author thought semen or flesh only contained various solids and fluids and that out of them, through respiration, the various parts of an animal develop in different stages. The question is, what does the author mean when he says that "each" (hekaston) goes as like to like. Does the hekaston refer to each of the fluids, such as blood, bile, phlegm, and water, mentioned in §3 of On Semen, or does it refer to bodily parts as Aristotle seems to have thought. In the former case, whatever the difficulties, there is genuine room for development, whereas on Aristotle's interpretation, development is simply magnification and revelation of what is already contained in the semen.

Aristotle's interpretation should not seem unfair because in §3 of On Semen, the Hippocratic author repeatedly asserts that veins and cords lead "from every part (morion) of the body" into the penis. These cords are filled with the various fluids I mentioned. Due to friction, the fluids are churned up, become foamy, which finally leads to the secretion of semen. The implication here is that these fluids contain portions of fluids and solids from the various bodily parts, such as a bit of fluid from the eyes as well as a bit from other moist parts, and a bit of solid from the ears as well as a bit of solid from the feet, and so on. This is how (L) seems to be used in the above quote. Due to respiration, these various portions of fluids and solids all go to their appointed places (hōrēn idiēn) from which they came. This idea of various portions of fluids and solids each being designated as portions from specific bodily parts and locations justifies to a certain extent Aristotle's literal interpretation. Aristotle reads pangenesis as the theory that the semen comes from all the parts of the body, ignoring the intermediate link of fluids and solids. Nevertheless, since our Hippocratic author designated portions of these fluids and solids to certain specific bodily parts and locations, the difference does not rule out of court the literal character of Aristotle's interpretation of (P). He has simplified the theory, but he has not distorted it. These slightly different versions of (P) do not affect Aristotle's criticism of (L) in Book II. I now turn to these criticisms.

One of Aristotle's most fundamental criticisms is that he does not see how a simple intake of air can be the sole cause of development. The most that can be said for such an intake, if that much, is that it is the cause of hardness, softness, brittleness, and the like, but it cannot be the cause of the _logos_ which turns these into flesh and bone and other uniform parts. For this the form-giving movement of the male is needed (II, 1, 734b3-6). The same holds, of course, for the uniform parts. Respiration might be the efficient cause, but it can never play the role of what Aristotle would call the formal cause. What is it that makes each portion of fluid or solid go to its appointed place?

This cannot be explained by an appeal to (L). For in that case, all the flesh would end up in one pile or location and all the bones in another, which is what happens in the example of the bladder into which air is piped.

> The differentiation of the parts does not happen, as some (_tines_) suppose, because like is naturally carried to like. For, among others, the theory has these difficulties. If one accepts this cause, it means that each of the uniform parts will come to be separately, for example, bones and sinew by themselves, and so will flesh. But the real cause why each of the parts comes to be is that the residue of the female is potentially (_dunamei_) such as the animal naturally grows to be; and the parts are present potentially (_dunamei_), but not actually (_energeiai_). (740b13-22)

One of the other difficulties is that our Hippocratic author has misrepresented the order of development. He believes that the uniform parts develop before the non-uniform ones, while Aristotle thought they developed simultaneously (734b27). Note how Aristotle consciously rejects (L) and replaces it with an explanation based on his own potential-actual distinction. The material of the female, which in this case is comparable to the material mixture of the two semen, does not really and actually contain the parts of an animal. Rather, they somehow grow out of it naturally. I return to this in later sections.

A similar contrast between (L) and the potential-actual distinction is made a few chapters later. Again the idea of spatial relocation of the constituents of the embryo is attacked.

> And what some (_tines_) physiologers say, namely, that 'things are naturally carried to their like,' must be taken to mean not that the parts are moved by changing their position, but that, while they remain at rest, they are altered with respect to softness, hardness, color, and other differentiating features of the uniform parts; they become actually (_energeiai_) what earlier they had been potentially (_dunamei_). (741b7-15)

This passage and the preceding quote show, I think, that Aristotle consciously replaced the like-like maxim with a principle of explanation based on his own potential-actual distinction. I indicated earlier that this "replacement of paradigms" in biology is a part of a much larger revolution in the philosophy of nature, a revolution which Aristotle precipitated with the introduction of his new potential-actual distinction as a principle of explanation. For the moment my thesis is that this replacement did take place in biology and that Aristotle was keenly aware of the shortcomings of (L) in an account of development. In light of these defects, he offered his own potential-actual distinction as an improvement.

I have also argued that the specific, historical appeal to (L) which Aristotle rejected is the one made by our Hippocratic author in _On the Nature of the Child_. We must remember that there is a very close link between _On Semen_ and _On the Nature of the Child_.[1] The latter takes off where the former left off and it presupposes the pangenesis theory expounded in that treatise. I have already given ample evidence that in Book I of the _G. of A._, Aristotle attacks the theory expounded in _On Semen_. Since he attacks a version of (L) in Book II of _G. of A._, it is reasonable to suppose that it is the version found in _On the Nature of the Child_ which, as I said, is so intimately connected with _On Semen_. Then, too, of the records we possess, the use of (L) to explain the development of an animal is nowhere as clear as in _On the Nature of the Child_. No other viable historical source is available. Furthermore, at several places in Book II, Aristotle refers to "certain people" (_tines_), as he does for instance in the above quotes. Each time, the views mentioned can be found in this Hippocratic treatise.[2] Although the evidence is, of course, not conclusive, it would seem that Aristotle's use of *tines* or *tines tōn physikōn* is his way of referring to the school of Hippocrates generally, or to the author of these two treatises, specifically. Finally, I should mention the fact that almost all the topics discussed in *On the Nature of the Child* are discussed by Aristotle in *Book II (and Book IV) of the G. of A.*[3]

Until another plausible alternative is suggested, or this one shown to be mistaken, I conclude that Aristotle wrote large parts of Book Two of the *G. of A.* with the argument from *On the Nature of the Child* in mind, just as he wrote *Book One* with *On Semen* in mind. The dialectical discussion that started in the first book is continued into the second.

2. *The internal motivation*

The failure of (L) is not the only reason why in Book II of the *G. of A.* Aristotle makes use of the potential-actual distinction. In addition, he was faced with an internal dilemma, one created by certain basic tenets of his own philosophy. A discussion of this dilemma will lead us into the crucial passages where the distinction is introduced and applied to the problem of development.

The form-matter hypothesis is a bold conjecture. It transfers the form-matter distinction from the realm of artifacts, where it is quite helpful, to the question of what the role of the sexes is in the generation of animals. It does not seem that the sexes make similar contributions. First, this is what the defenders of (P) supposed when they said that both sexes secrete semen from all their bodily parts. But (P) has been shown to be false. So, perhaps the sexes do not contribute equally. Second, there is also the fact that the male and the female do have different reproductive organs which suggests that they perform different functions in the generative process.[4] Why, if they perform similar functions, would they have dissimilar organs with which to perform those functions? Third, Aristotle has been able to point to cases of generation where the male did not contribute any material and to others where the female would insert part of herself into the male, all of which suggests that the role of the male is not the transference of any material. Since he obviously does play a crucial role, it is best to suppose that he contributes the form of the offspring. Against this background, the analogy between the carpenter and the male, though a bold one, does not seem too farfetched.

There is one fundamental difficulty with the analogy, however, and Aristotle hits it hard in Book II. It is easy to see how the carpenter, being in constant contact with his material, is able to impose a form upon that material. This requirement of contact between agent and patient holds for any kind of action and passion (Generation and Corruption I, 6, 322b23-24; 323aff). Furthermore, only those things can be in contact "which possess a definite magnitude and position and which have their extremities together" (323a5). Clearly, a carpenter, while at work fashioning his material, is in contact with that material in the way here specified. But how does the male in the procreative act meet these requirements? Long after the male has withdrawn and gone his own separate way, the development of the offspring starts and continues without the need for further male assistance. It seems that Aristotle's bold conjecture is too bold for his own theories about action and passion. How can there be action and passion where there is no contact? This is the problem that lies behind the admission, early in Book II, that "concerning these matters [of generation] there is a big problem as to how any plant or animal is generated out of the embryo (spermatos - 733b24).[5] The topic for discussion is "not out of what but by what the parts are formed" (732b32). If the male imposes the form and is the formal cause, what could possibly be the efficient cause? To put it simply, how does the male do it?

The problem results from Aristotle's acceptance of the popular view that the semen is that from which all the parts of an animal develop (734a7), together with his belief that there must always be contact between the agent (the male) and the patient (the female). The conflict arises because nothing can be seen to be in contact with the embryo during development, nothing external nor anything internal. Yet it would seem that the motive cause, which makes the contact between agent and patient, has to be either external or internal to the embryo (733b33).

The details of this dilemma are as follows; the motive cause cannot be something external, for:

> It would seem unreasonable that something (ti) external fashions each of them, whether the viscera or the other parts, because if it is not in contact it cannot cause movement, and if it does not cause

> movement nothing can be affected
> on account of it. (734a2-5)

The external motive cause must affect the embryo. But to do this, it must move the embryo and hence be in contact with it. If the motive is indeed something external, then we must be able to observe that something is in contact with the embryo during its development. Since we do not, the motive cause cannot be (attached to) something external. It must, therefore, be something internal.

In this case, there are two possibilities. Either the (internal) cause is a natural part of the organism, or it is a foreign and separate, though internal, part of it (734a5).

It cannot be a separate and foreign part. For if it were, it must either remain till the animal is fully developed or be destroyed during the process of development. It does not remain to the end for when the organism is fully developed, all the parts of it are natural (organic) parts, and no foreign elements are present (735a8). Nor is it likely that this foreign element went out of existence before the process of development was completed. For suppose it fashioned the heart and then disappeared, leaving it to the heart to develop the other parts? In this case, we have two motive causes, the foreign element and the heart. Now the heart does not disappear after it has performed its creative function, so why should we suppose the foreign element did disappear after its creative task was finished? Is it not more likely than not that all motive causes behave in the same way; they either all disappear or they do not.

The only choice left is that the motive cause is an internal and integral--and not a foreign--part of the organism (734a13). If so, this motive, natural part must have caused the development of the (other) parts, either simultaneously or successively. Observation precludes the first alternative. But even if the development took place successively, we face the problem of one bodily part causing another bodily part to be formed. Aristotle rejects this possibility on the grounds that if the heart, let us say, was the motive cause of the other parts, it would have to have the shape of those other parts (734a30). But clearly the heart does not have the shape of a liver. Therefore, it is absurd to suppose that one natural, bodily part could be the cause of all the others. It cannot even be the cause of any one of them.

There is, however, an even more fundamental objection which does not depend upon the claim that the cause must have the shape of its effect. The idea that one part is the cause of the others is unworkable because it conflicts with the accepted opinion about what the semen is. It is that from which all the parts develop. The semen then is formed first of all by the generating parent. Now if one bodily part is the cause of development, then that one part must have been ready made in the semen from the beginning, and it was not formed by the semen. For if it was, then the semen was the cause of its development and that part is no longer the sole efficient cause. All of this conflicts with the accepted belief, supported by observation, that all the parts of an animal develop from the semen. That is what is meant by semen, namely, that from which (all) the parts of an animal develop (724a17).

The conclusion is that "that which fashions the parts is not present within the embryo. It is not an internal and integral part, nor is it a foreign part. But it cannot be something external either, yet it must be either of these?" (734b4) What, then, is the motive cause of the development of animals?

The form-matter hypothesis by itself does not suffice to answer this question. Just like the Hippocratic author supplemented (P) with (L), so Aristotle supplements his form-matter hypothesis with his potential-actual distinction.

Aristotle's first move in resolving the predicament is to reexamine one of the premises:

> Perhaps one of the things we said is not that simple, i.e., why is it impossible for the parts to be formed by something external. In one sense it is possible, but in another sense it is not. (734b5-7)

He has second thoughts about the internal-external dichotomy laid down earlier and now introduces[6] the potential-actual distinction by drawing an analogy between the process of generation and the phenomenon of some amazing automata:

> It is possible that A causes B to move, and B C, very much like the amazing

> automata. For the parts, even though at rest, somehow have a power (dunamis) that when something external sets the first one of them in motion immediately the one next to it becomes actual (energeiai). Just like in the case of the automata, in one way that which causes the movement is not now any more in contact, having, however, been so, so also that by which the semen came to be or which fashioned the semen, while having been in contact with it. And in another way the movement (kinēsis) is present within, just like the activity of building causes the house to be. (734b10-17)[7]

The point of the analogy is not that bodily part A causes bodily part B to be and take shape. That possibility was rejected at 734a33. Rather, it is the idea of automatic development which provides a compromise between the internal-external dichotomy that created the predicament.

The analogy teaches us not to view the efficient cause as a thing or a tode ti but as a movement (kinēsis - b17) or power (dunamis - b12). That Aristotle had this contrast in mind is clear from the conclusion he draws at 734b18-20. The predicament has been solved because we now don't need to look anymore for some internal or external part that is in continuous contact with the embryo. Why not suppose instead that generation occurs in the way the automata start to move, namely, automatically after the first push has been given.

This analogy does not solve all of Aristotle's problems. He is still open to the charge that this movement which moves the automata is a random movement which does not contain or transmit any sort of logos or eidos which would make the automata move in a specific direction. I am here thinking of a row of domino stones that is knocked over. They all fall, but there is no way of telling precisely how and where they will fall. In what way, then, is the movement and power here introduced an improvement over the respiratory process which the Hippocratic author saw as the efficient or motive cause?

Aristotle gets at this problem when he introduces, right after the automata passage, a new principle (archē) of explanation. This principle combines the potential-actual distinction with the form-matter hypothesis:

> [To explain] how then each [part] comes to be, it is necessary to start from here; first we should accept the principle (archēn) that all things that come to be in nature or by art come to be by that which is actually (energeiai) such and from that which is potentially (dunamei) such. (734b20-22)[8]

(Note how the "by that which" is ambiguous between the motive and formal cause. I return to this later.)

According to this principle which I shall call (M), any O comes to be by that which is actually O and from that which is potentially O. It would seem that (M) plays the same role in Aristotle's attempt to solve the problem of development as (L) did in the attempt of our Hippocratic author and that (M) replaces (L). Though something like this is no doubt true, the word "replacement" is a bit strong here. For both (L) and (M) are expressions of the same, fundamental belief shared by all Greeks of Aristotle's day, that nothing can come from nothing. From this belief, the Greeks derived the maxim that a cause must resemble its effect. For if it does not, it must be supposed that the effect arose ex nihilo, which is absurd.[9] The defenders of (P) and the users of (L) interpreted this requirement of resemblance literally. If an animal comes from the semen, then they supposed the semen must have had in it all the parts of that animal. Like must have developed out of like. Aristotle rejected this literal interpretation of the resemblance requirement. Instead of making one single causal factor exactly like the effect, he spread the requirement out over several causal factors. This is what he does in (M). The "that by which" must be potentially like it. While the requirement of resemblance has been preserved, hopefully some absurdities have been avoided.

Right after the introduction of (M), Aristotle applies it to the problem of development:

> Now the semen (sperma) is such a thing and it has such a movement (kinēsin) and principle (archēn) that as soon as the movement has stopped each of the parts comes to be and gets besouled. (734b23-25)

Since Aristotle has been defending the form-matter hypothesis, it would seem unobjectionable if we identified the *sperma* here as the female matter and the *archēn* as the male's contribution. In the next two sections, I defend and explain this interpretation and application of the potential-actual distinction. Though this is both the natural and the correct interpretation, something must be said about both identifications. In Section B, I shall defend my identification of the "that from which" with the female matter against a recent and different interpretation. In Section C, I shall give a brief explanation of what Aristotle meant by "that by which," since, as I mentioned earlier, this phrase is ambiguous between the formal and motive cases.

B. The "that from which" defended

The word *sperma* is not easy to translate. Although Aristotle has a different word for the male's residue (*gonē* - 727a7 and 27), for the female residue (*katamenia* - 727a29 and b32), and for the product of these two (*kuēma* - 728b34), he, at different times, has used *sperma* to refer to each of these items. At 728a28, he uses it to refer to the same male residue referred to at a26 with the word *gonē*. (See also 739a18.) At 728a7, he says that the female contribution is a kind of *sperma*, though not as pure as that of the male. (See also 737a28.) And in numerous places, the best translation for *sperma* is indeed the "embryo." For instance, throughout my discussion of Aristotle's internal predicament about the motive cause (Section A), I spoke of the embryo. I pictured Aristotle as being puzzled about the motive cause because nothing internal or external could be seen to be in contact with the embryo. The problem resulted from Aristotle's acceptance of the popular belief that the *sperma* is that from which all the parts of an animal develop and his requirement that the agent be in contact with the patient. The present passage comes shortly after this discussion. There is, therefore, some warrant for saying that the "that from which" an animal comes to be, as specified in (M), is to be identified with the embryo.

I rejectthis reading because the entire context has been (731b1-732a10 and 733b25) and is that of the theory of the four causes. The very formulation of (M) witnesses to this fact. Aristotle has been concerned to nail down the motive cause of the generative process and, as he is increasingly successful in doing so (cf., the automata analogy), the material cause also receives a clearer focus. The introduction of (M) has pushed that clarity a step further. So while it was all right to translate sperma with "the embryo" in the earlier part of the discussion, it is misleading to do so at 734b23. This means that strictly speaking, the "that from which" an animal comes to be is the katamenia supplied by the female. That, and not the embryo, is the matter of an animal, and it is that which is potentially an animal. No other interpretation fits the theory of the four causes. To say that an animal comes from the embryo is not correct if the "from" introduces the matter of the animal or that which is potentially an animal.

This view of the matter of an animal conflicts with a recent proposal made by Barrington Jones in his paper "Aristotle's Introduction of Matter."[10] Jones believes that the embryo is the matter of an animal and that from which an animal comes to be. He argues this as part of his general thesis about the relativity of matter. I shall briefly explain Jones' thesis and then argue that my interpretation of matter as being the residue of the female fits in better with Jones' overall thesis than does his own view of matter as being the embryo. I am not concerned with whether or not the relativity thesis itself is correct.

Jones' paper is a refinement of and in one sense an improvement upon the views Charlton expresses in his commentary on Physics I and II.[11] In the appendix to this work, Charlton has successfully (let us suppose) dealt with Aristotle's use of the phrase protē hulē--successful in the sense that he was able to maintain his view that the matter of something is what it is made of, in spite of the "prime matter" references. This already makes matter relative for there are lots of stuffs from which things are made, and there is no primeval stuff that underlies all the stuffs which we know about, like marble, earth, or flesh and bone. As Charlton sees it, the matter of something is what it is made of, and that stuff is what is potentially that thing and from which that thing came to be.

But Charlton is puzzled by Aristotle's remark in Physics A, 7, 190b4: "For there is always something which underlies[12] from which what comes to be comes to be, e.g., plants and animals from a seed (ek spermatos)."[13] This remark does not fit Charlton's view of matter:

> Flesh and bone do not become or turn into a dog; they are not, under any description, a terminus a quo of the

> change to a dog, but rather a *terminus ad quem*. It is the seed which is the *terminus a quo*.... There is thus a serious gap in Aristotle's argument. (76-77)[14]

The gap is that, in this case, that from which the product is made (flesh and bone) is not the same as that from which it came to be (the semen). So which is the matter of an animal?

Jones' paper is in part designed to solve Charlton's puzzle.[15] He does so by arguing (and I grant him the arguments of Sections II-IV) that the matter of something is not the stuff from which it is made but rather the particular *piece* of that stuff which was turned into the product. Thus Jones draws a distinction between that from which something is made and that from which it comes to be. The latter and not the former is identified with the matter. The matter of a statue is the unshaped piece of marble which the sculptor used and not the stuff, marble. It is that particular piece which was potentially a statue. Consequently,

> 'Matter' is an antecedently existing individual which comes to be the product and...the product comes to be 'from' this, not in the sense that it is made from it, or made from a type of stuff, but where 'from' is used in a purely chronological sense. Thus the product comes to be 'from' the matter solely in that the matter is something that existed before the product.[16]

This passage makes it clear that "matter" is a purely relative concept even more so than for Charlton. Now matter is not just relative to the kind of stuff used to make the product; it is relative to every piece of one kind of stuff that is turned into any kind of product.

In Section V (from the end of which the above quote was taken), Jones applies this analysis to the *Physics* passage which had caused Charlton to stumble. He argues that

> There is no reason why the animal should not come to be from the embryo in just the same way that the statue comes to be

> from what it is in fact said to be from, the unshaped piece of bronze. Just as that piece does not remain throughout the change, so neither does the embryo survive the birth of the animal that it comes to be.[17]

And since the embryo is a piece of flesh and bone,[18] Charlton's puzzle has been solved. The matter of animal is not what it is made of (flesh and bone) but it is a particular piece of flesh and bone, namely, the embryo (sperma). Thus, Charlton was almost right. All he needed was the distinction between "that from which" something is made, and "that from which" it comes to be. What is potentially an animal is the embryo and it is the embryo from which an animal comes to be and that is a piece of flesh and bone.

I shall give five arguments against Jones' identification of "that from which" with the embryo. (None of them concerns the relativity thesis itself, which I have accepted for the sake of the argument and with which I am, in fact, sympathetic.) In raising these objections, I hope to show that my identification of "that from which" with the female residue fits better with Jones' own thesis than does his own identification. I further hope to show that there was more to Charlton's puzzlement than met Jones' eye.

First, to view the embryo as the matter of an animal does not fit well with the form-matter hypothesis. Aristotle never changed his mind on that hypothesis and in a work as late as Metaphysics VIII, he clearly states that the material cause of a man is the menstrual blood (katamenia - 4, 1044a35).

Second, Jones' view that the matter is a particular piece of stuff is supported by Aristotle's explanation of the cause of multiple births and of the redundancy of parts (G. of A., IV). The difference is that the stuff is menstrual blood and not flesh and bone, but the talk about pieces and chunks of matter fits right in. Aristotle claims that in each procreative act, the female secretes a certain limited amount of seminal material (hulēs spermatikēs - 722a), which is not unlimited, not too big (pleon), and not too small (elathou). The offspring cannot come to be out of just any amount (eks hoposesoun toi plethei - a28). The male, too, emits a certain amount of semen with just enough power to give shape to that piece of female material (demiourgoun - 740b25 and 771b22).

Nature, which never does anything in vain, is so arranged that both partners secrete the required amount--no more and no less--either just enough for one offspring or just enough for two or for more. Once in a while, a little bit will get left over. In that case, there is a redundancy of parts. So while Jones is right about the matter of an animal being a certain piece of stuff and not simply the stuff itself, he is wrong on the kind of stuff it is. It is not flesh and bone (which, incidentally, is two kinds of stuff and not one), but blood. In the Parts of Animals, Aristotle refers to blood as the hulē of all the parts of an animal, which is distributed to those parts through the blood vessels. (III, 5, 668a4)

Third, in viewing blood as the matter of an animal, we can do justice to Charlton's suggestion (ignored by Jones) that flesh and bone are more of a terminus ad quem than an a quo. Certainly, blood is a terminus a quo; we can justifiably say that an animal comes from a piece of blood without feeling the oddness of saying that it comes from a piece of flesh and bone. An animal does not come from flesh and bone, not even a piece of it, as Jones suggests; it is (among other things) a piece of flesh and bone. Aristotle nowhere says that the embryo is a piece of flesh and bone. In fact, such a claim would conflict with his view that all the parts of an animal, including flesh and bone, develop from the embryo.

Fourth, Jones is a little bothered by the disanalogy between saying that a statue comes from the unshaped piece of marble and that an animal comes from--what? He answers the question with the point of Aristotle's that when we don't have a name for the privation, we may say that something comes from what underlies (but does not remain), i.e., the piece of stuff. "Thus," says Jones, "we seem to be left with no alternative but to say that it is indeed the embryo that comes to be an animal from being an embryo."[19] An easy solution would be to say that the animal comes to be from the undeveloped embryo. And one even more fitting with the above arguments would be to say that an animal comes to be from the unshaped female matter. This is perfectly consistent with and even suggested by Aristotle's comparison of the male and his semen with the carpenter and his tools. The female matter is like the unshaped pile of wood. The art-nature analogy fits better than Jones thinks it does.

Finally, I return to the possible translations of the word sperma. It is used in the Physics passage, which Charlton translates as follows:

> There is always something which underlies, out of which the thing comes to be, as plants and animals come to be out of seed (ek spermatos). (A, 7, 19025)

It is unclear whether Charlton means by "seed" the female menstrual blood or the embryo. His commentary does not clear up this ambiguity. (Clearly, he cannot mean the male's semen as that from which an animal comes.) Jones definitely wants to translate ek spermatos with "from an embryo." He appeals to certain passages in the G. of A. where sperma can indeed refer to the product of the two contributions, i.e., the embryo. So, it is possible to read the passage the way Jones wants to read it. This reading would allow more clearly for the analogy between plants and animals. But we don't have to read it this way. We can strain the analogy a bit and hold firmly to the theory of the four causes, especially since this is a discussion of that very theory.

A second passage that plays a role in the dispute is Metaphysics IX, 6, 1049a14, where Aristotle says that the sperma is not yet potentially a man, but only after it has been changed into something else, i.e., an embryo. Jones interprets Aristotle as insisting here that "although the animal can be said to come from the semen (gonē) properly speaking, this is not what it comes to be from."[20] We are now faced with two modes of speaking properly, the one I introduced in terms of speaking in the context of the four causes and the one Jones has in mind here.

The mode in the Metaphysics has to do with Aristotle's distinction between movements and actualizations introduced a few chapters earlier. Properly speaking, x is potentially y if x does not need to undergo any changes before it can become y. An adult, for instance (to use an example from the De Anima) is ready to become a general. But x is not potentially y if x first needs to undergo changes before it can become y. A boy first has to become an adult before he can be a general. Similarly, the sperma, in the sense of the female matter, is not potentially an animal properly speaking. It first has to be fertilized by the male's semen (gonē). After that, it is potentially an animal in the proper sense of the word potential. This distinction may well lie behind Aristotle's observation at 729a34 that the menstrual blood is like _prōtēn hulēn_.[21] The embryo would be the second matter. Depending on the context, one way is a more proper way of speaking than the other.

Just like there are two senses of "being potentially," the one that involves a change and the one that does not, so there are now two senses in which x is the matter of y. Depending on which kind of matter we have in mind, the matter of an animal will be either the embryo or the female matter. What kind of potentiality did Aristotle have in mind in _Physics_ A? None, I suggest, for by that time he had _in all_ likelihood not yet developed that distinction in conjunction with the theory of the four causes. The only reference to the distinction in _Physics_ A is at the end of chapter eight and that may _well be a_ later addition.[22] So, let us reverse the question: What kind of matter did Aristotle have in mind in _Physics_ A? I suggest the matter that fits the theory _of the_ four causes. And in that context, the proper way of speaking is to say that the _katamenia_ is the matter of an animal.

I conclude that in terms of the theory of the four causes, i.e., properly speaking, the "that from which" an animal comes to be and its matter is the menstrual blood supplied by the female. And it is that matter too which is potentially an animal. The relativity of matter can be maintained (supposing it deserves to be maintained) without doing violence to the theory of the four causes.

C. The "that by which" explained

In Section A, I pointed out that Aristotle introduces the potential-actual distinction with a reference to some amazing automata. These automata operate the way our domino stones do; once the first one is set off, the other ones automatically follow. Aristotle uses this example to show how at one point in the development the motive cause can be external and in contact with the embryo while later it has become a completely internalized movement. The notion of automatic development is Aristotle's solution to the problems raised by his own requirement that in every case of movement, an agent must be in contact with the patient. In the case of the development of animals, no such contact can be observed. Therefore, the efficient cause must be internal to the embryo. And since it was already determined that the male provides the form or the movement (= the form-matter hypothesis), the internal movement inside the embryo is the one contributed by the male parent. The example of the automata helps explain how that is possible.

But a shift occurs when at 734b20, Aristotle introduces what I have called maxim (M). He says that in order to explain the development of animals, we should start with the principle that any O which comes into being comes from that which is potentially O and by that which is actually O. The "that by which" here refers not (just) to the efficient cause which up to this point has been the focus of the discussion and which was stressed in the automata analogy. Rather, the "that by which" must be the formal cause. For how can a movement or motion be (actually) like a thing or a product such as an animal?

The formal cause in this case is the male generating parent, "who is actually (entelecheiai) what the material from which the offspring comes is potentially (dunamei)" (734b35 and 735a5). We have already seen that the material referred to here is the female menstrual blood upon which the male somehow imposes a shape or form. This somehow is what needs to be clarified. Granted that the male himself is the formal cause and does indeed have the same eidos as the offspring, and granted that the movement of his semen is the efficient cause, how does the latter get the former across?

Let us return to the analogy between the male and the carpenter for a moment (I, 22, 730b9-25). Aristotle compares the semen of the male to the tools of the carpenter. Just as the tools do not become part of the product, neither does the semen become part of the offspring. Its bulk evaporates (737a15). The eidos of the male is compared to the idea the carpenter has in his head or to the blueprint in front of him. The carpenter intentionally moves his tools in such a way that the desired order or shape is imposed upon the material. Analogously, the semen is the medium by which the male shapes the female matter. What is missing, however, is the intentional character of the process. Clearly, the male does not look at his own eidos and then moves his semen in such a way that the same eidos is imposed on the matter. How, then, does the semen perform its function?

Aristotle argues that the male semen can be and is a carrier of the male form because it is a residue of final nourishment or blood.

> Because the semen (*spermatos*) is a residue it is moved *with the* same movement (*kinēsin tēn autēn*) with which the *body grows through* the distribution of final nourishment, when it has entered the uterus it sets and gives the same movement to the residue of the female with which it itself happened to be moved. (737a18-22)

The implication of this passage is that the final nourishment or blood of the male contains the *logos* (formula) or *eidos* (form) of the male. The semen, which is a residue of that blood, therefore also contains that same formula or form. The bulk of this semen comes into contact with the female matter, implants this movement, carrying the male's form in that matter, and then evaporates. After this, the female matter automatically develops into the bodily parts of an animal according to the script of the implanted movement.

It is tempting to view Aristotle's talk about the same movement (a19 and 22) as referring to one continuous, unbroken motion or flow of energy from the male to the semen and on to the female matter. This, however, is only a minor aspect of the truth. Aristotle does view continuity, being the absence of intermittent periods of rest, as a criterion for counting a certain movement as one and the same movement (*Physics* V, 4, 228b5-10). Clearly, the movement involved in copulating is one and continuous.[23] There are, however, two other criteria that warrant discussion because they help explain what Aristotle means by the same movement, which in turn will help us understand what he means by the "that by which" phrase as used in (M). For something to be one and the same movement, it must also be one in respect of that in which it moves and, furthermore, the that which is moved must also be one and the same.

The that "in which" does not refer to the locus of the movement, as might be expected, but to the *eidos* of the movement. Every movement must be from something (*eks hon*) and to something (*eis ho* - *Physics* V, 1, 224b). This whence and whither constitutes the form of a movement (224b26).[24] The form of walking would be the distance covered. The form in the category of quality might be gradations between the coldness of the object and its consequent warmness. In generation, it is the "distance" between non-existence and existence (224b10). In *Physics* III, Aristotle points out that "that which causes movement shall always bring with it a certain characteristic (eidos ti)...which shall be the principle and cause of movement (21, 202a10). This is true of every category. Thus, in the category of substance, that which causes movement (e.g., the male) carries with it the male's eidos and imposes it upon a new piece of material (e.g., the female residue). In the case of our example, the whence is not the male and the whither is not the female, but these terms refer to the "distance" between the unshaped katamenia and the shaped or developed katamenia. The movement, in the case of generation, is that which brings about the change from the one state of privation to the other state of possessing the characteristic in question. As Terry Penner puts it, movement "is what intervenes between something potentially having a predicate and actually having it."[25] Now the movement in question is no doubt one in form because the whence and the whither are the same. That is to say, the entire movement of copulating must be viewed as bringing about the shaping of the katamenia. Every movement must have its whence and its whither. It must have a product, so to say. This is why the word "movement" is a better translation for Aristotle's kinēsis than our word "motion." The word motion does not get across the idea that every movement must have an eidos, a whence and whither. The product is missing in motion.

This does not mean that there is no locus for the movement. There is. For Aristotle, the locus is invariably the patient or that which is moved. So, for there to be one movement, that which is moved must also be one and the same thing (227b25). In terms of our example, it means that the piece of female matter must be one and the same for there to be one movement and not two. Thus, if Coriscus impregnated two women, there would be two movements. Even in the case of multiple births, one movement is split up into several according to the number of pieces of material involved.

We, of course, might be tempted to locate the movement in the male and his semen as well as in the female matter. But Aristotle does not do this. He recognizes that there are two potentialities being actualized, the one of the male agent, and the one of the female patient, but he argues that there is in this case only one movement, namely, the one in the patient (Physics III, 3, 202a12ff).

The "that by which" in maxim (M) refers, therefore, to the male partner, who is actually like the offspring. He is the formal cause, or the blueprint. He succeeds in informing the female matter with his form because his semen is a residue of final nourishment or blood which is that which makes the bodily parts grow and thus a likely cause of what makes them come to be in the first place. The semen serves as tool and transmits the movement from male agent to female patient. Even though the movement is in the male, the semen and the female matter, it is still one and the same movement. The reason is that the whence and the whither are the same, namely, from the undeveloped state of the material to its developed state. Furthermore, the patient and the time period are also one and the same.

We must note that in the above explanation of the shift from the motive cause, as stressed in the example of the automata, to the formal cause, as stressed in (M), I relied heavily on 737a18-22, where Aristotle links the same movement (kinēsin tēn autēn) to final nourishment and blood. In other words, Aristotle's explanation of how the male imposes his eidos on the female matter is, for lack of better terms, a mechanistic or materialistic one. Blood is the underlying cause of why the offspring ends up having the eidos of the male. The curious thing is that this is the same reason Aristotle gives to explain the fact that the female matter is potentially the bodily parts of an animal. At I, 730a15, for instance, he refers to the female contribution as the nourishment or the matter. Aristotle openly acknowledges that the female menstrual blood is blood (I, 726b36); it just is not as well concocted as the male's semen. My point is that Aristotle seems to trace both the male and female roles to blood, which reduces the radical character and impact of the form-matter hypothesis. If the male can do what he does because his semen is a form of blood, and if the female can do what she does because of the same reason, then what is left of the form-matter hypothesis?

The answer is that Aristotle seems to have two views of the male semen. In Book I, he argues that it is concocted blood. But in Book II, 3, when he discusses the problem of development, he says that semen is made up of pneuma and water, making no mention of the bloodlike character of semen, which figured so prominently in Book I. While the former description of semen plays down the radical character of the form-matter hypothesis, the latter description preserves that radical character. Aristotle never tires of saying that the parts of an animal can only be such parts if they are alive and have souls (e.g., 734a, 734b25ff). He argues that the key soul for an animal, i.e., the sensitive one, is contributed to the offspring by the male only. And the _pneuma_ is the carrier of this soul principle. This means that the male is responsible for the fact that the offspring has a nose, ears, and eyes, etc. The male and not the female is the cause of the fact that out of the female matter develops a living, unified, and besouled animal. Aristotle had criticized the Hippocratic author on the grounds that the pangenesis theory does not explain how all the different parts of the bodies of the parents that are collected in the embryo do, in fact, become unified and form a living animal. In tracing both the male and the female contributions to blood as a residue and as that from which the parts come to be, Aristotle has made some improvement. But the essential question is still left unanswered. That is the question of the source and origin of the order or _logos_ of the bodily parts and their being alive and able to perform the functions that living beings perform.

Since Aristotle believed that the presence of soul(s) is essential to the generation of a living animal, he owes it to the Hippocratic author to explain how an animal gets its soul(s). Just as in his explanation of the acquisition of bodily parts, so here too Aristotle relies heavily on his newly introduced potential-actual distinction.

D. The Acquisition of Soul

In Section A, I quoted a long passage from §6 of _On the Nature of the Child_. The first line of that passage reads as follows: "As the flesh grows, it is differentiated by air (_pneumatos_) and each kind of substance in it goes to its own kind." In that first section, I stressed the use made of the like-goes-to-like maxim so evident in this passage. Partly by way of reinforcing my claim that Aristotle took our Hippocratic author as his major opponent, I now wish to draw attention to the claim that air is the efficient cause of the development of an animal.

The first paragraph of this little treatise makes it clear whose air this is.

> Then it [the semen] forms air partly because it is in a warm place and partly because the mother breathes. When it is filled with air, the air in the middle of

> the semen makes itself a passage to the outside through which passage it escapes. (§1)

It is the *pneuma* of the embryo itself primarily, and secondarily that of the mother. Aristotle too gave a role to *pneuma* in the generative process, but he objected to the account just quoted. He says that

> The parts of the animals are differentiated by air (*pneumati*), not, however, by the *pneuma* of the mother nor by that of the [offspring] itself, as some natural philosophers say (*tines tōn physikōn* - 741b37-42a)[26]

For Aristotle, too, *pneuma* is the efficient cause of development, but he thinks it is the *pneuma* of the male and not that of the embryo or of the female. He also has a different view of the nature of this *pneuma*.

For the Hippocratic author, the air involved is simply one of the four basic elements, the other three being earth, fire, and water. Not so for Aristotle. He believes the male semen to be made up of water and *pneuma*, which is hot air (736a2). This air is not one of the four elements, but more divine than they are (736b32). It is the sublunary analogue of ether (737a). This heat is not that of ordinary fire, which does not generate life, but it is like the heat of the sun, which is known to generate life (737a1-3). The presence of the *pneuma* or hot air accounts for the foamy appearance of the semen when first ejected (735b11). After a while, the earthy, watery bulk of the semen evaporates, and all that it left is this life-giving, form-giving *pneuma* (737a15). In a backward reference to the predicament discussed in Section A, Aristotle observes that, because the bulk has been evaporated, "it is not necessary to look for its leaving the female externally or to find it as an (internal) part, once it has been set and shaped" (737a14). What the male contributes is simply this hot, divine, fertile air.

It is not entirely clear whether Aristotle thought the soul to be identical with this *pneuma* or very closely connected with it. I think the latter is the best interpretation. It is true that at 737a9, Aristotle refers to this hot air as the soul-principle (*psychikēs archēs*). And he says of this principle that it is emitted together with the bulk (*sōma*) of the male semen. Sometimes this soul-principle is separable from that bulk and sometimes not. This line of thought is open to the claim that the soul and the *pneuma*, which clearly is the other item emitted with the bulk, are one and the same. Still, Aristotle calls it a soul-*principle*, and not merely soul *per se*. That is, the *pneuma* is the moving principle and carrier of the soul, which fits much better with what Aristotle said a few lines earlier.

A few lines earlier, he had said that the power (*dunamis*) of every soul seems to have something to do with (*kekoinōnēkenai*), a physical substance (*sōmatos*) different from the elements (736b30). This substance turns out to be the *pneuma* or hot air I have already described. But "having something to do with" and "being one and the same thing" clearly are not one and the same thing. There are, therefore, three things to be distinguished: 1) the *dunamis* of each soul; 2) the *pneuma* with which each power seems to be connected; and 3) the bulk and watery part of the semen which serves as a physical basis for the *pneuma*, which, though physical itself, is quite a rarified substance, not at all like the ordinary elements. Hence, the interdependence of soul and body goes both ways. Just like a bodily part needs to be besouled to be the part of a living animal, so the soul depends upon the *pneuma* which depends upon the watery bulk of the semen. When this bulk evaporates, this *pneuma* presumably attaches itself to the female matter and is internalized by it, thus allowing the soul to get to work on the automatic development.

This dependence of the soul on the body receives even clearer expression in the *De Anima*, written at about the same time as the *G. of A.*[27] In the *De Anima*, Aristotle compares the relationship between the soul and the body to that of vision to the eye. Vision is the *ousia* or being of the eye, and the eye is the matter of the vision. This same line of thought, says Aristotle, must be carried over to the soul-body relation as a whole (412b33). He then compares the soul and the body, which together make up a living animal, to the faculty of sight and the pupil, which together make up an eye that can see. And for a body to be so besouled (simply?) means that it can perform the various functions of an animal; it can think, sense, move, or be at rest, and grow or decay (413a22). The soul, by way of these enabling functions, differentiates what lives from what does not live. Since living things perform these functions--and that is how we tell they are alive--the soul is the activity (*entelecheia*) of a body that has organs required for such activities. Aristotle defines the soul as "the first actualization of a natural body possessing organs" (412b5). It is the first actualization because even an animal that is asleep can do what an animal does.[28] When it actually does these things, it is in its second actualization. The mere possession of the organs and the concomitant ability counts as the first actualization. Says Aristotle, "It is therefore quite clear that neither the soul, nor a certain part of it, if it has parts, can be separated from the body. For some (souls) the actualization of the parts is their actualization" (413a4). After a remark like this it should come as no surprise to us that when Aristotle sets out to explain how and when an animal acquires its various souls that that explanation depends a lot upon his explanation of how an animal acquires its bodily parts.

After Aristotle has explained the acquisition of bodily parts in terms of the potential-actual distinction, he applies the same sort of reasoning to the problem of the acquisition of souls. Obviously, since the semen is the only physical link between generations, the origin of the soul is intimately tied up with the kind of thing semen is. The following passage makes this connection evident:

> Whether the semen has soul or not? The same argument [applies which applied] with respect to the parts of the body. For 1) neither shall soul be in anything other than that of which it is soul and 2) neither shall any part be such in more than name unless it has soul in it, like the eye of a person who has died. It is clear that (3a) it has soul and (3b) is soul potentially (dunamei) (735a5-10)[29]

In 2), Aristotle affirms that all the parts of an animal need to have soul just to be such parts. Since the semen is part of an animal, it too must (3a) have soul and be alive. In this respect, the semen or fetation is like any other part of an animal. Although it is not entirely clear from this passage itself, it seems that 1) must be the basis of (3b). Perhaps the connection is this: the soul which is in the parts that are formed cannot be in any other part nor can it come from any other part or thing. If we suppose that a part is already formed at time t, we must also suppose that it has soul at t. A part cannot receive soul after it has been formed. Rather it has soul right from the beginning. It seems reasonable, then, to suppose that the soul comes from the same source as the parts. Since the embryo contains all the parts potentially, it must also contain soul potentially. Viewed in this way, the semen has soul potentially in general, i.e., it potentially must have the life of the parts that it potentially contains.

There is, however, a more specific way in which the embryo has soul potentially. It also has each sort of soul, nutritive, sensitive and rational soul, potentially. To say that the embryo has potentially soul in general means in fact to claim, for Aristotle at least, that the embryo has these three sorts of soul potentially since they constitute the life of an animal.

There are degrees of potentiality, "for it [the embryo] itself may be nearer to or further from that which it is capable of being potentially, just like the sleeping geometrician is further removed than the one who is awake, and that one than the one who is theorizing" (735a9).[30] Aristotle does not precisely indicate with which geometrician we are to compare the nutritive soul. He makes it clear, though, that the nutritive soul develops before the other two souls. The situation is, therefore, the following: the embryo has all three souls potentially, but it has the nutritive soul closer to being actual and that soul develops earlier than the other two.

The idea that the embryo has the nutritive soul earlier is, according to Aristotle, clear from observation and is agreed upon by most people. Some embryos have the nutritive soul immediately, while others have it only potentially, so to say, until they are separated from the parent at which time they have that soul actually. At an early stage, "all things of this sort [fetations, etc.] have the life of a plant" (736b13), and, hence, have a nutritive soul. This is not a matter of dispute since "the semens and fetations of animals are just as much alive as plants are and up to a point they are fertile (736a33). Now if we compare these fetations that live and grow and have been separated with those that are not yet separated but live off their parent, we can say that the latter have the nutritive soul potentially during the time before they are separated. After they have been separated, they have it actually and grow by themselves. (In humans, the heart is the organ through which the nutritive function works, and this organ is, therefore, fashioned first.) Aristotle says, "We must posit that semens and fetations which are not yet separated possess it [the nutritive soul] potentially, though not actually, i.e., not until they begin to draw nourishment to themselves and perform the function of the nutritive soul" (736b9). Having thus accustomed his readers to the notion that these fetations have the nutritive soul, at least for some time, potentially, Aristotle suggests that the embryo has the other two souls in the same way.

Of the other two souls, the sensitive and the rational soul, the same sort of thing can be said as was said of the nutritive soul, "for it is necessary that all [souls] must exist potentially before they exist actually" (736b15). The whole passage reads as follows:

> It is clear that we should say the same sort of thing about the sensitive soul. For it is necessary that all of them exist potentially (dunamei) before they exist actually (energeiai). It is further necessary that either all souls, if they do not exist beforehand, develop inside (the embryo) or that they all exist beforehand, or that some do and some do not. And if they develop inside (the embryo) then they do so either in the matter, not having entered by way of the semen of the male, or they have entered the matter by way of the semen. And if they enter by way of the male semen then either all have been formed in the male semen, or none have, or some have and some have not.

I take it that the series of options spelled out in this passage is meant to support the initial claim that the embryo has all the souls potentially before it has them actually. That is, the various souls develop from the embryo for the same reason as do the parts, i.e., because it has them potentially. Step by step, Aristotle eliminates the different possibilities until the claim has been established that the embryo has all the souls potentially. The potential-actual distinction is an essential ingredient of Aristotle's explanation of how the souls--and hence living animals as over against mere parts--develop from the embryo.

The following steps reflect the options Aristotle has spelled out. Either:

1) All the souls exist beforehand, or
2) They all develop in the embryo.

If the latter, then either

3) They develop out of the female matter, or
4) They develop in the embryo, having entered the embryo by way of the male semen.

If by way of the male semen, then either

5) All of them came to the male semen from the outside, or
6) None of them did (and they all developed in the semen of the male), or
7) Some did and some did not come to the male from the outside.

Although Aristotle does not explicitly say so, it would seem that what 1) means with respect to the souls is what (P) meant with respect to the bodily parts. Aristotle took (P) to mean that all the parts of the body exist in the semen because the semen was drawn from all the parts of the body. In (L), the possibility is offered that all souls similarly preexist in a ready-made manner. Against this possibility, Aristotle introduces another maxim, (N)

> For of as many principles as the activity is physical it is clear that these [principles] cannot exist without a body, for example, there can be no walking without feet (736b22).

This maxim is a form of the more general statement that every function needs to have an instrument (716a24; 765b37). If the function is a physical activity such as walking, then the function cannot exist unless the bodily instrument exists. In like manner, the activities expressive of soul are such that the performance of these activities requires (with the exception of the rational soul) bodily parts as their instruments. Without those bodily parts, such as eyes, ears, etc., the sentient soul cannot exist and neither can the nutritive soul exist without the appropriate bodily organ. To claim, then, that the souls exist before the embryo has developed, and that they are, therefore, not developed from the embryo amounts to claiming that the bodily parts which serve as the instruments of those souls exist beforehand, which is absurd. Therefore, 1) does not explain how the animal receives its souls. This leaves 2) through 7).

The same principle, (N), also rules out two other possibilities, namely 5) and 6). The first states that all the souls came to the male semen from the outside. This is impossible for at least two of the three souls, the nutritive and the sensitive, which cannot exist in separation from certain bodily parts through which they perform their function (736b24). For these two souls to enter the male semen, their respective bodily organs would somehow have to attach themselves or be added to the semen which is absurd. Therefore, it is not the case that all souls enter the male semen from the outside. But the rational soul is an exception. Since "its activity has nothing in common with physical activity" (736b28), it could have entered the male semen from the outside. And if it did, then 6) must be rejected, for we could not claim for certain--as does 6)--that none of the souls entered the male semen from the outside. This leaves possibilities 3), 4), and 7).

According to 3), all three souls develop on account of the female matter. The phenomenon of wind eggs shows that this cannot be true. By itself, without the intervention of the male, such an egg cannot produce an animal able to perform the functions of the sensitive soul (741a10ff). This also shows that 4) is not a real possibility. According to this supposition, all three souls enter the matter by way of the male semen. But wind eggs do grow by themselves and are alive before male intervention; therefore, the nutritive soul does not enter by way of the male semen, and 4) must be rejected. This leave 7) to be discussed.

In this case, it is supposed that some did and some did not enter the male semen from the outside. Aristotle does not reject 7) because it is possible that the rational soul entered the male semen from the outside. Since the rational soul does not require a specific bodily organ to function, of all the souls, it is the only one that may have an origin independent of the origin of physical bodily parts. This is not true of the sensitive soul. It must be developed from the semen the way the bodily parts required for its operation develop. Hence, one may and one cannot enter the semen from the outside, and 7) presents a real possibility.

The upshot is that each soul has a separate origin and that the semen has all the souls potentially. The rational soul is the only exception in that it may have entered the male semen ready-made from the outside. The female matter has the nutritive soul potentially before the embryo is separated and nourishes itself and before it comes to have that nutritive soul actually. It has the sensitive soul potentially simply because, and in the same way as it has, the bodily parts. Any principles, the activity of which is physical, cannot exist apart from the appropriate bodily organs. The semen of the male which carries the movement that guides the development of the parts also is the transmitter of the soul without which those parts would not be the parts of an animal.

E. Conclusion

I began this chapter by saying that it was not enough for Aristotle to raise a series of objections to the pangenesis theory as I discussed them in Chapter III. One of Aristotle's major objections was that pangenesis does not explain how a living unified animal develops from the embryo. If (P) is the sole principle of explanation, should the constituents of semen not have a random order rather than the well-ordered parts of an animal that develop out of it.

In Section A, I pointed out that the Hippocratic author supplemented (P) with the maxim that "like naturally goes to like." I discussed Aristotle's objections to (L) and pointed out that he consciously replaced the appeal to (L) with an explanation based on his own potential-actual distinction. In addition to the defects of (L), there are also good reasons within Aristotle's own perspective that favour the adoption of the new distinction. Aristotle believes that in every case of movement, the agent has to be in contact with the patient. The problem is that while development is a case of movement, or better, involves movement, no agent can be seen to be in contact with the embryo during the time of development. As a result, Aristotle gave up a strict interpretation of the contact requirement and came to view the movement and the efficient cause of development as internalized by, and being internal to, the embryo. He felt the development might be an automatic one as in the case of the amazing automata.

Just as the Hippocratic author had to supplement (P) with (L), so Aristotle supplemented his form-matter hypothesis with the potential-actual distinction. I defended the view that the female matter is what is potentially the new animal. It is from her residue that the parts of a new animal automatically develop and come to be. But first, this development needs to be set off by the male parent who is actually like the offspring. He transmits his form and soul by way of his semen, which has in it the same movement as does the male himself, a movement representative of the male eidos.

The discussion of this chapter has been almost exclusively exegetical. I have argued that Aristotle replaces (L) with (M); I have defended and explained the respective parts of the potential-actual distinction, and I have shown how Aristotle used the distinction both to explain how an animal against its bodily parts, as well as how it gains its souls, all the while sticking close to the text. Even without the evaluative comments, for which the reader will have to wait till chapter VI, the points I have made are important for the thesis I am defending. In the preceding chapters, I have shown how Aristotle is engaged in Book I of the G. of A. in a dialectical discussion with a certain Hippocratic author. I have now shown that this discussion is continued in Book II. There is a textual parallelism between the discussion of the nature of semen in On Semen, and the matter of subsequent development discussed in On the Nature of the Child, and a similar division between what for us are Books I and II of the G. of A. There further is a philosophical continuity because Aristotle cannot offer his form-matter hypothesis as yielding a superior explanation without offering his own solutions to the problems he saw with an appeal to (P) and (L). There can be no doubt that Aristotle wrote large parts of Book II with the Hippocratic treatises I mentioned in mind. He continued in Book II what he started in Book I. In the next chapter, I discuss how he brought it to a close in Book IV.

FOOTNOTES

[1]Littré, the editor and translator of the French edition of the Hippocratic Corpus (Oevres Complètes D' Hippocrate, Paris, J. B. Baillière, 1851, Vol. VII, p. 470-542), numbered all the paragraphs from On Semen to On the Nature of the Child, and into the fourth book of On Diseases consecutively. He believed that all these sections originally formed one treatise on embryology. While there is general agreement among scholars that the first two of these originally formed one treatise, this agreement is not carried over to the third. See Wilhelm Kahlenberg, "Die Zeitliche Reinenfolge Der Schriften peri gonēs, peri physios paidou und peri nousōn 4 und ihre Zusammenhörigkeit," Hermes, 83, 1955, 252-256.

[2]I, 17, 721b13 states that some accept the pangenesis theory. (See On Semen, §3.) II, 4, 740a13 talks about those like Democritus who beleive that the external parts are formed first. (See On the Nature of the Child, §6, as quoted in the text.) At 740b14, (M) and (L) are contrasted. (See same quote.) The same contrast is made at 741b9. II, 6, 741a1 mentions those who believe that the pneuma of the creature itself is the cause of differentiation. (See same quote.) IV, 3, 769a7 refers to the pangenesis theory again and how it is used to explain the phenomenon of resemblance. (See On Semen, §8.) Many of these correlations are footnoted by Peck in the Loeb edition of the G. of A.

[3]I shall briefly mention the main topic discussed in each paragraph of On the Nature of the Child and place behind it in parentheses the place where Aristotle discusses the same topic(s) in the G. of A. §1:breathing is the efficient cause of the formation of the first membrane. (III, 1, 733b24 initiates a discussion of the efficient cause of development.) 2: test of the woman who was six days pregnant and lost the embryo by jumping up and down (no reference). 3: growth by blood of mother (II, 6). 4: formation of flesh, navel first (II, 4 and 6). 5: flesh again, membranes and cavities (II, 6). 6: differentiation of parts, the like-like maxim (II, 6). 7: length of time needed for conception (IV, 10). 8: development of nails, fingers, toes (II, 6). 9: hair and grey hairs--castration changes appearance (V, 3; I, 2, 716 b5). 10: milk (IV, 8). 11: plants (no discussion, but see I, 1, 716a, and scattered references like 739b35, 740a26, b11, b6). 12-16: transplants (no extended discussion, but see 723b16ff). 17: arrangement in womb (II, 6). 18: navel first, test of 20 eggs (II, 6; heart first, but no mention of egg test). 19: delivery is due

to a lack of nourishment or matter (IV, 10, and see especially the equation of nourishment and matter at I, 19, 727b19). 20: the causes of twins and greater number of offspring and the origin of the sexes (IV, 1-4).

[4]See Book I, chapter 2.

[5]In the present passage, I have translated sperma by "embryo" because Aristotle seems to be wondering how a living organism develops from whatever it is that it is normally thought to come from, which is the semen or embryo. He is not now questioning conventional wisdom about the material cause of development (cf. also I, 17, 721b6). He simply wonders what the motive cause of it all is. Later, however, at 734b20-22, he does use the "that from which" phrase to refer exclusively to the material (residue) contributed by the female and "that by which" phrase refers to the male's contribution.

[6]He also used it at 734a30, but there it was not yet crucial to his argument, since he goes on to make his point in another way that does not rely on this distinction. Now, however, the distinction does become crucial to the argument.

[7]There are a few other passages where Aristotle refers to these automata: G. of A. II, 5, 741b9; Generation and Corruption, II.

[8]Other versions of this principle occur in Metaphysics N, 1088b15;L,2,1069b15; K, 1065b5; A, 7, 1017b; E, 2, 1026a33; Physics III, 200b26; Generation and Corruption 317b15 and 23; De Anima II, 414a25; De Coelo III, 302a6.

[9]cf., Metaphysics K, 1062b25.

[10] The Philosophical Review, October 1974.

[11] W. Charlton, Aristotele's Physics I, II (Oxford: Clarendon Press, 1970).

[12] "But the underlying thing is not a third factor over and above the opposites: it is the same thing as one of the opposites, viz. that from which the change takes place, but under a different description," Charlton, p. 73. In other words, "underlies" does not mean "remains unchanged." (Generation and Corruption I, 319b21-31)

[13] Jones' translation, p. 489.

[14] Charlton, p. 76.

[15] Jones, p. 475.

[16] Jones, p. 493.

[17] Jones, p. 490.

[18] Jones, p. 492.

[19] Jones, p. 492.

[20] Jones, p. 489. Aristotle does not use gonē here, which word refers specifically to the male contribution. Such a reference is wholly out of place in this context. Aristotle uses sperma, which can be ambiguous between embryo or female menstrual blood.

[21] Charlton does not know what to make of this reference (p. 130). Heinz Happ in Hyle Studien Zum Aristotelischen Materia - Begriff (Berlin-New York: Walter de Gruyter, 1971) believes this is a reference to an "absolute qualityless substrate" (p. 552). My interpretation is consistent both with the relativity of matter thesis, which I have provisionally accepted, and with the two kinds of potentiality Aristotle held to during the last part of his career. This distinction is maintained in the De Anima and there are references to the De Anima in the G. of A.: II, 3, 736a35-b1; 4, 415a23-25; V, k, 779b21-26; 7, 786b23-25. It is therefore possible that these two senses of potentiality, spelled out in Metaphysics IX, and used in the De Anima, lie behind the use of protē hulē at G. of A. 729a34.

[22] See A. P. Bos in Een Onderzoek naar de Kosmologie van Aristoteles (Amsterdam: Filosophisch Instituut V.U., 1971), p. 130.

[23] Generation may involve a movement, but it is not a movement. See Physics V, 225a.

[24] For a good discussion of this point, see Terry Penner's "Verbs and the Identity of Actions - A Philosophical Exercise in the Interpretation of Aristotle" in Ryle: A Collection of Critical Essays, editors Oscar P. Wood and George Pitcher (Garden City, New York: Doubleday and Company, 1970), sections II and III. The present discussion was triggered by Penner's paper and is obviously (e.g., eidos of walking being the distance

covered) indebted to it.

[25] Penner, p. 429.

[26] See Note 2.

[27] See Note 23.

[28] This conflicts with Henry B. Veatch's explanation of what Aristotle means by "first actualization" (Aristotle: A Contemporary Appreciation, Indiana University Press, Bloomington, 1974). Veatch thinks that "first" refers to the substantial change something must undergo to become an animal, the soul "being that which brings about a primary or initial [i.e., substantial] actualization of something that is alive, or if you will, of a living body able to perform such life functions as nutrition, growth, and reproduction. Once such a thing as a living substance has thus come into being, all of such subsequent actions represent changes in a substance that is already existing; as such they are in the proper sense of the term accidental changes which the existing substance undergoes" (p. 64). Two things are wrong here. First, the idea of "first actualization" does not refer to the coming to be of an animal, but to its mere ability to perform the various functions indicative of the presence of life. As such, Aristotle compares the case of the soul to that of a sleeping geometer. Second, when an animal actually does any of these things and is engaged in its second actualization, that is not really a case of accidental change because it is not (properly speaking) a case of change at all; it ignores Aristotle's distinction between movement and activity (cf., De Anima III, 7, 431a4; Physics III, 2, 201b32; Metaphysics IX, 6; XI 1066a18; Physics VIII, 257b7; Nichomachean Ethics VII, 15, 1154b27; and Rhetoric III, 11, 1412a9).

[29] Nuyens takes this passage to support his view that in the G. of A., Aristotle held the same view about the relation of the soul to the body as he did in the De Anima, namely, that the soul is the entelechy, form, or "life-force" of the body and is not localized in any one bodily organ, rather, it is present in all bodily party equally (Ontwikkelingsmomenten in de Zielkunde van Aristoteles, Nÿmegen: Decker & Van de Vegt N.V., 1939, p. 237ff.). I do not wish to dispute this view itself (see, for instance, II, 4, 738b25-27), but most of the passages to which Nuyens refers, including this one, do not support his thesis in any direct and straightforward manner. The main point of this and other passages (I, 19, 726b22-24; II, 1, 734b

24-27) is that in order for a bodily part of an animal to be such a part, it must be "besouled." The contrast here is not explictly with the view that the soul dwells in one bodily organ, as Nuyens repeatedly suggests, but simply with a corpse. Nuyens misinterprets the second line of this quote. He takes Aristotle to say that before he can answer the question of whether the semen has soul, we must ask the same question regarding the other parts of the body (p. 239). The Greek reads as follows: "O autos logos kai peri ton morion." I have translated "autos logos" to "the same argument" and believe that it refers back to the argument discussed in Section A. Nuyens, on the other hand, believes that the question about whether or not the semen has soul is no more important than when it is asked regarding other bodily parts; he thus generalizes the question and loses sight of the crucial role played by semen in Aristotle's account of development. Having broadened the question unduly, he fails to appreciate how Aristotle, even in his final period struggled with the question of the soul's dependence upon the body. That question was not shelved when he apparently gave up his view that the soul depends upon one bodily organ.

[30] Marjorie Grene thinks that the explanation of the acquisition of bodily parts and of soul, which I have been discussing in this chapter, is typical of the sort of explanations Aristotle gives as a scientist (A Portrait of Aristotle, Chapter III, Section 6). But she misses the most important point of Aristotle's entire procedure; she claims that "it is the distinction between grades of potentiality and actuality, applied to the nature of the efficient cause, that has provided the solution" (p. 102) to the problems of development. It is not the grades which make the explanation typical, but the very use of the potential-actual distinction itself. She ignores the novelty of the distinction in the philosophy of Aristotle and does so in the context of a discussion of his typical approach as a scientist.

CHAPTER V

THE THEORY OF INHERITANCE

By the end of this chapter, we shall have arrived at the end of the dialectical dispute I have been discussing. Chapter VI is an evaluation of the dispute. Like the preceding one, this chapter, too, is largely exegetical in character. I argue that in Book I of the G. of A., Aristotle did not give an adequate response to the argument from resemblance on behalf of (P). He, therefore, takes up the matter of resemblance between parents and offspring again in Book IV, at which time he gives a much more elaborate and satisfactory explanation. This is important, because without a good explanation of the phenomena of resemblance between both of the parents and the offspring, Aristotle will not have accounted for the strongest piece of evidence for (P).

The theory of inheritance discussed in this chapter also contains Aristotle's answer to the problem of sex differentiation. This was the third major objection Aristotle raised against the pangenesis theory. Thus, having responded to the problem of development in Book II and to the other two problems in Book IV, the dispute can be considered closed.

A. The Unfinished Argument of Book I

In §8 of On Semen, we read this:

> And in the semen itself there is some drawn from each part of the body, both the woman's and the man's; and weak semen from weak parts, and strong from strong parts. There is necessarily a corresponding result in the child. From whatever part of the man's body more is contributed to the semen than from the woman's, in the offspring that part is a better likeness of the father; from whatever part more is contributed from the woman, that part in the offspring is a better likeness of the mother.

In chapter 17 of Book I, Aristotle gives a fairly accurate report of this argument in these words:

> Besides these [mutilations] there are [positive] resemblances with respect to the parents. For offspring comes to be which resembles [the parents] not just with respect to the whole body, but also part for part. If then the cause of the resemblance of the whole is that the semen comes from the whole, then the cause [of the resemblance] of the parts would also be that something comes from each of the parts. (721b21-24)

It seems that Aristotle was aware of what sort of defense could be given for (P). He knew to what evidence our Hippocratic author appealed. Formalized, the argument might go like this:

1) There is an <u>overall</u> resemblance between parents and offspring.

2) Semen is the only physical link between generations. It comes from the parents and the offspring comes from it.

Therefore,

3) Semen must be the cause of this overall resemblance, but

4) There are also cases where parents and offspring resemble each other in their <u>parts</u>.

Now if we suppose that

5) Similar effects must have similar causes

We may conclude that

6) Semen is also the cause of these detailed cases of resemblance;

And the best explanation for how this can be is to suppose that

7) In the semen, there is a portion of <u>each</u> bodily part (the presence of which determines the resemblance in the part of the offspring).

My charge is that in Book I, Aristotle never adequately responds to this argument.

Obviously, our Hippocratic author saw (P) as an explanation, not just of the overall resemblance between members of the same species, but especially of the specific or detailed cases of resemblance between one of the parents and the offspring. From whichever part of either parent the biggest portion of semen is drawn, that part will in the offspring resemble the parental part from which the large portion was drawn. The analogous part of the other parent is outquantified, so to speak. In chapter III, I discussed some of Aristotle's criticisms of this argument from resemblance. Aristotle's objections do indeed show that (P) leaves a great deal to be desired as an explanation of resemblance. To suppose that the semen is literally made up of parts from all the bodily parts leads to all sorts of difficulties. We can see, though, from the passage I quoted that the Hippocratic author did have some response to, for instance, the charge that the offspring must be bisexual. He would say that it all depends upon what portion from what parent was outquantified. In Book I, Aristotle ignores this aspect of (P) and simply lists what he thinks are some of the absurd consequences that follow from a literal reading of (P). Furthermore, while he ignores this dominant-recessive aspect, he gives an explanation of resemblance which fails to come to grips with the detailed cases of resemblance (P) was designed to solve.

Twice in Book I, Aristotle links his own explanation of resemblance to his redefinition of semen as a residue of final nourishment, i.e., blood. He did not think semen was the kind of stuff which had just come <u>from</u> all the bodily parts, for nature always discards <u>such</u> residues while providing ducts for the semen to flow through and a place where it collects (725a3). Semen seems to be a natural part of the body, a residue of nourishment which is about to be distributed <u>to</u> all the bodily parts (725a24-28, 726b20-24). In the following passage, Aristotle gives this constitution of semen as a residue of blood as the explanation of resemblance.

> And [on account of the fact that semen is concocted blood] it is understandable that children come to be like their parents. For that which is distributed to the parts resembles what has been left over.

> Thus the semen is the semen of the hand or of the face or of the whole animal. And what each of these is actually, such the semen is potentially, whether with respect to its own bulk or because it has a certain dunamis within itself. (726b13-19)

This link between semen and concocted blood explains the overall resemblance between parents and offspring, but no provision is made for the detailed cases of resemblance, such as when a girl has her father's crooked nose. Yet it is precisely this kind of case that (P) was designed to explain. What is missing is a mechanism which tells us how the semen as blood transfers individual characteristics from either parent to the offspring. Our Hippocratic author provided for this when he claimed that the semen was made up of parts from all the bodily parts of both parents, but that some parts dominate or outquantify others. Aristotle does not give such a mechanism until Book IV, and until then he has not met the argument from resemblance.

There is still a second defect in his answer in Book I. We must remember that the semen referred to in the passage just quoted is the semen of the male and not that of the female. She does not discharge semen because she is not hot enough (727a). But Aristotle does say that the female residue, too, is a form of blood; it just is not as well concocted as the male's residue. It turns out, then, that the resemblance between the parents and the offspring is entirely due to the male's semen as concocted blood. This means that the degree of concoction is what counts and not the mere fact that semen is a form of blood. For if the latter were the case, then the female too should be able to provide part of the form. Unless further elaboration is provided, this clearly is a weak link in the argument, especially because the bloodlike character of the female's residue is far more obvious than that of the male's residue, the bloodlike character of which is only revealed under very special circumstances (726b9). This further elaboration is provided in Book IV, where we learn that in Book I, Aristotle seems to have had only the ideal cases of resemblance in mind.

No wonder, then, that in Book IV Aristotle returns to the matter of resemblance. After he gives his own theory of inheritance in the third chapter, he briefly remarks on how wrong other interpretations are. This is one of the explanations he rejects:

> Certain natural philosophers had different views of these matters. They gave different explanations of why the offspring are formed similar or dissimilar [to their parents]. They speak of the cause in two ways. For some say that the offspring will resemble that more from whichever most of the semen came, similarly the whole after the whole and part after part, as if the semen is drawn from each of the parts. If an equal amount comes from both then the offspring resembles neither. But if this is false and the semen is not drawn from the whole then it is clear that the cause which they gave for similarity and dissimilarity is not [the real cause]. (769a7-15)

This sounds remarkably like what our Hippocratic author said in the passage I quoted from On Semen. Evidently, Aristotle offers his own theory of inheritance in Book IV against the background of the failure of (P). Our Hippocratic author is still his major opponent. What is more, I argue below that Aristotle borrowed the essentials of his own theory from the very opponent he is trying to refute.

B. The Theory of Inheritance

Both before and after he states his theory of inheritance, Aristotle gives a rundown of the phenomena the theory is designed to explain. The shorter of the two passages reads as follows:

> It has now been stated on account of what cause females and males come to be, why some after their parents, female after female and male after male, and others the reverse, the female after the father and the male after the mother, and generally why some resemble their ancestors and some after none and these things with respect to the whole body and with respect to each of the parts. (769a)

Every one of these phenomena was the basis for one or more objections Aristotle raised against (P) in Chapter 18 of Book I. He repeats some of these objections in Book IV:

> While those who maintain that the female and the male come to be because more or less semen comes from either male or female have no way of showing how a female can take after the father and a male the mother. For it is impossible for more [semen] to come from both. Further, on account of what cause does the offspring take mostly after the ancestors, even after distant ones? Surely no part of the semen has come from them. (769a19-26)

The phrase "more or less semen" is yet another reference to the belief of our Hippocratic author that the quantity of semen determines after which (part of which) parent the offspring shall take. Aristotle offered his own theory of inheritance as a better explanation of those phenomena which (P) was designed, but failed, to explain.

Aristotle begins by extending the meaning of dunamis far beyond its ordinary usage. It now becomes a technical term adopted for theoretical purposes. Suppose Coriscus engages in a procreative act. As a generating parent, we can view him in at least four ways: as an animal, as a human being, as a man, and as the individual, Coriscus, he in fact is. In each case, we have a different set of characteristics in mind. When we view him as an animal, we probably think of him as being a living creature with nutritive and sensitive soul and made up of flesh and bone. When we view him as a human being, we might add that he has an intellective soul and whatever feature differentiates man from the other animals such as being a featherless biped. As a male, he was given reproductive organs and the ability to concoct and set the material supplied by the female. Finally, when we view him as the individual Coriscus, we add all his peculiar characteristics such as a high forehead and a cheerful disposition. I use the latter example even though in his discussion, Aristotle seems to have only physical characteristics in mind. He seems to view the set of characteristics that make up the individual Coriscus as the set of the dunameis of all his bodily parts (cf., 768b1-4, 767b1). But the schema Aristotle gives can be used for mental as well as for physical characteristics and we need not prejudice the case by supposing that Aristotle thought all mental phenomena reducible to physical ones. Whatever the case, his schema fits either way. There are, therefore, in the procreative act of Coriscus four _dunameis_ present, each _dunamis_ comprising the set of characteristics that makes him an animal, a human being, a man, and an individual respectively.

Before I quote the passage I have in mind, let me state the main points Aristotle makes. 1) He lists the four _dunameis_ of which I spoke. 2) He says that the movements (_kinēseis_) which cause the parts to develop from the embryo come from these _dunameis_. (A movement is not identical with its _dunamis_, but seems to proceed from it. 3) He claims that some _dunameis_ are only potentially (_dunamei_) present. (By implication, then, these movements are also present potentially.) 4) The _dunameis_ that stand closer to and are more peculiar and characteristic of Coriscus are stronger than those that stand further away and are less peculiar. The closer ones, therefore, dominate in the process of development. Coriscus, as a particular individual (_tode ti_), is _ousia_ and has the most (power of) being. (Note the use of _ousia_ here as in the _Categories_ and not as in _Metaphysics Z_.) Here is the passage:

> Each _dunamis_ I mean in this way. That which generates is not only a male, but also a certain kind (_toion_) of male, for example Coriscus or Socrates, and it is not merely Coriscus but also a human being. And in this manner some (_dunameis_) are closer to and some further from that which generates, _qua_ such that can generate, but not according to what it accidentally is, for example, if he is a good scholar or the neighbor of someone. And always with respect to generation the peculiar and that according to the individual is strongest. For Coriscus is both a human being and an animal; but the human being is more peculiar of him than the animal. Both the individual and the genus generate, but the individual more so. For this is the _ousia_. And that which is generated also comes to be a certain kind of thing (_poion ti_) but at the same time a certain particular (_tode ti_), and this is the _ousia_. Therefore the movements in the semen proceed from the _dunameis_

> of all such, and potentially (dunamei) also from the ancestors, but more likely from that which each time is closer to the [dunameis] of a certain individual. And by individual I mean Coriscus and Socrates. (767b23ff.)

Aristotle has gone far beyond the usual sense of the term dunamis, even with respect to his potential-actual distinction. For now certain dunameis are said to be present potentially (dunamei). Here dunamis has the technical sense described above. A dunamis is the "power" behind a given set of characteristics, be it the characteristics of an animal, a man, or an individual. What kinds of things are these dunameis, or what is their ontological status?

In the next chapter, I answer this question and argue that these dunameis are theoretical entities for Aristotle. They are postulates of a theory. It may be of interest to note now that the author of On Ancient Medicine,[1] who may be--but need not be--the author of On Semen and On the Nature of the Child, made a move very similar to what Aristotle does here. This author took exception to the view that all diseases could be traced to the imbalance of the hot, the cold, the dry, and the moist. He did not think those elements were the most crucial factors in matters of health and disease. Instead, he argued that there are all kinds of hot: the hot astringent things, the hot insipid things, and so on. The nature of man consists of many powers. In fact, every kind of food has its own specific effect on the body and thus has its own characteristic power. There are as many powers as there are effects. What Empedocles and other natural philosophers have said about the basic components of men pertains no more to medicine than to painting (XX, 10). The author recommends a thorough-going empirical approach to the question of how many and what sorts of dunameis there are in a man. He believes that there are in a man not just the conventional elements but "salt and bitter, sweet and acid, astringent and insipid, and a vast number of other things possessing dunameis of all sorts, both in number and strength" (XIV, 35).

Aristotle does a similar thing with respect to the problem of inheritance. He postulates a dunamis for each effect or part. The parts of the body are, so to speak, symptoms of the underlying dunameis. Aristotle does not seem to view these dunameis as properties of other yet more basic entities. Rather these dunameis are the most basic components of a man and are described as having certain *kinēseis* which transmit their characteristic "natures." Looked at in this way, Aristotle's talk about both *dunameis* and *kinēseis* is not redundant.

It is fairly obvious from the entire trend in the *G. of A.* that these *dunameis* are not thought of as being material parts or chunks of matter. In Book I, Aristotle argued that the *dunamis*, and not the bulk of the semen, is the cause of resemblance (I, 21). The same separation occurs in Book II when Aristotle resolves the predicament about the efficient cause by separating the *dunamis* and the *kinēsis* from the agent that initially contributed that *dunamis* to the female matter. So now, too, in Book IV, Aristotle postulates these *dunameis* without grounding them in some physically existent matter. This is surprising in a way because he had argued in Book I that blood as the final nourishment is the cause of resemblance. So why not argue now that these *dunameis* are somehow present in blood. One of the reasons why Aristotle may not have done this is that in Book II, he has given us a different description of semen as made up out of *pneuma* and water. Aristotle would not have had to claim that the blood *qua* blood is the cause of resemblance, but he might have claimed that the blood has certain as yet unidentified *dunameis* which cause resemblance. Nevertheless, in either case, Aristotle has reached a higher level of abstraction than the defenders of (P). They believed that the semen itself came from all the parts of the body. Aristotle claims that the semen consists of all the *dunameis* of all the bodily parts. This difference is important because it allows Aristotle to explain the resemblances between offspring and remote ancestors. Both Hans Meyer[2] and Erna Lesky[3] make the point that Aristotle was the first thinker to tackle this problem of the resemblance between offspring and remote ancestors. There is no material link of the semen in these cases of resemblances. Nevertheless, Aristotle argues, we must assume that the *dunameis* of these ancestors are present in the semen of later generations. If we do not assume this, we cannot account for these cases of resemblance. Let us now return to Coriscus and his *dunameis*.

Aristotle purposely picked Coriscus and Socrates as male examples. He has not given up his form-matter hypothesis. He reiterates in Book IV that the fundamental difference between male and female lies in the ability to concoct. The male has the ability to concoct, to cause to take shape, and to discharge semen "which has the principle (archēn) of the form (eidous)." By "principle," Aristotle meant "the proximate motive principle" (tēn kinousan protēn - 765b10-15). The male supplies the first principle of movement because he is hotter and thus better able to concoct and give shape to the material (765b17). No other defense is offered. Aristotle clearly viewed as the ideal case an instance where Coriscus, a male, produces a boy that looks like himself in all respects (769a22, 768b8). He even thought of a female as a deformed male (737a28). So in Book IV, he does not let go of his form-matter hypothesis. What he does offer--and this he did not do in Book I--is an explanation of everything that deviates from the ideal case.

Suppose we represent the procreative act of Coriscus linearly as follows, starting with the dunameis that are furthest removed from the individual: 4) the dunamis of an animal (and its movement)→ 3) the dunamis of a human being (and its movement)→ 2) the dunamis of a male (and its movement)→ 1) the dunamis of Coriscus (and its movement)→ the offspring. What Aristotle has to explain is how anything ever deviates from this ideal case, how a girl is born, how a boy gets to look like his mother, how a child takes after long-gone ancestors, and how monstrosities are born that don't seem to resemble anything at all. To explain these deviations, Aristotle introduces three hypotheses; the three hypotheses are all introduced together:

> It is necessary to lay down three general hypotheses (hupotheseis), one we just spoke of, namely, that 1) of the movements some are potential and some are actual. And two others: 2) that which gets mastered (kratoumenon) departs into its opposite, 3) that which relapses (luomenon) goes into the movement next to it; if it relapses a little into the movement close by, if more, into one further removed. (768b4-10)

These three principles, together with the technical sense of dunamis and kinēsis already discussed, make up Aristotle's theory of inheritance. Let us refer to the second hypothesis as the mastery principle and to the third as the relapse principle. I mentioned earlier that Aristotle borrowed some of the key elements of his theory from our Hippocratic author. I had particularly in mind the idea that some movements overpower or master others. This notion is central to the explanation the Hippocratic author gives of why a male or female child is born and, by implication, since the semen comes from all the bodily parts, why a child resembles its father or mother. In §6 of On Semen, we read:

> The male sperm is stronger than the female for of necessity the male develops from the stronger sperm. And here is another point. If the sperm that comes from both is strong, a male is born, if weak, a female is born. Whichever prevails in quantity, that is what is born; for if the weak sperm is much more abundant than the strong, the strong is overpowered and being mixed with the weak is diverted to female; if the strong is more abundant than the weak, and the weak is overpowered, this is diverted to male.

The verb used for "prevail" and overpower is kratein which is the same verb Aristotle used in his formulation of the mastery principle. The difference between the two is that the Hippocratic author links the strength of the semen to its quantity or bulk, whereas Aristotle assigns the strength or weakness to his newly postulated dunameis and their movements. But the idea of being overpowered and then diverted to the opposite is the same in both. Aristotle has simply added two more hypotheses and raised the discussion to a more abstract level, thus avoiding the absurdities of a too-literal interpretation of the constituents of semen.

In the remainder of this chapter, I show how Aristotle's theory of inheritance succeeds where pangenesis was said to fail and at the same time I show how Aristotle handles the resemblance cases that deviate from what he has laid down as the norm. These two sets of problems overlap, and I make no attempt to keep them separate. The hypotheses I just introduced will be more fully explained as we go along.

At 769a, Aristotle summarizes the phenomena he had sought to explain, and I pointed out that each one of them served as a basis for an objection to pangenesis. They are: the differentiation of the sexes; the resemblance of either offspring to either parent; the resemblance with ancestors; and the birth of monstrosities.

The birth of a male resembling his father has already been explained. It is the Coriscus model. In this case, the *dunamis* of Coriscus, *qua* male and *qua* individual, is not overpowered. Applying the mastery principle to the male, *qua* male, Aristotle says at 766a19 "that when the *archē* (or *dunamis*) does not gain mastery...and is unable to reduce [the material] to its own proper form (*eis to idion eidos to hauton*), then it necessarily changes into its opposite," and a female is born. Of the movements (*kinēseis*) that constitute the semen, some are present actually (*energeiai*), such as the one of (male) generating parent and those of a general kind, such as the one of a man and of an animal. Others are present potentially (*dunamei*), such as the one of the female and those of the ancestors (*tou thēleos kai tōn progonōn* - 768a14). A female is born because the *dunamis* of the female, *qua* female, was not mastered by the *dunamis* of the male, *qua* male. That female *dunamis* is present all along potentially, but it reveals itself when the male *dunamis* lacks its customary heat. Then the potential movement of the female is actualized.

Aristotle has made a very important admission here. If the female, *qua* female also has a *dunamis* with its correlative movement, even if that *dunamis* is not normally operative, that suggests that she too can, in principle, supply (part of) the form (*eidos*) of the offspring. Furthermore, she not only can but, in practice, often does put her stamp on the offspring. The *dunamis* of the female, *qua* female, is *not* mastered as often as it is mastered by the *dunamis* of the male, *qua* male. In Book I, Aristotle defended what he considers to be the ideal case, while in Book IV, he brings the ideal in line with the real. The admission that the female, too, has her own *dunamis* *qua* female and also *qua* her ancestors is an important qualification of the form-matter hypothesis. I return to this point in the conclusion.

Next to be explained is the resemblance of a boy after his father and a girl after her mother. In the first case, the *dunamis* of Coriscus *qua* individual dominates completely. In the second case, the *dunamis* of Coriscus both *qua* male and *qua* individual are overpowered by the mother *qua* female and *qua* individual. When a boy looks completely like his mother, the *dunamis* of Coriscus *qua* individual but not *qua* male is overpowered. When the girl looks like her father, the *dunamis* of Coriscus *qua* male is overpowered, but his *dunamis qua* individual dominates. Usually, of course, children look a little like both of their parents. Aristotle accounts for this with the supposition that each part of the body also has a *dunamis* and its correlative movement. "For of parts some movements are actual and others potential, as was said often" (768b4). In whichever part the child looks like the father, the movement from that part of the father dominated in the generative process. Similarly for detailed resemblances to the mother.

Compare this explanation to the following passage from §8 of *On Semen*:

> There is of necessity a resemblance in some part to each of the parents, if indeed the sperm passes into the child from both bodies, and whichever [parent] contributes more to the likeness, i.e. [contributes] from more parts of the body, that is the one that the child resembles in more parts. It happens sometimes that a daughter born resembles the father in more parts than the mother and a boy born looks more like his mother than his father.

All Aristotle has done is to shift the source of the resemblance from the semen itself to the *dunameis* of the various parts. He objected to the explanation just quoted because he did not think semen came from all parts of the body. But by rejecting pangenesis, he has also rejected any physical grounding of his postulated *dunameis*, unless, of course, he grounds them in the blood, which he did not make much of an attempt to do.

The relapse principle is needed to explain resemblances between offspring and ancestors. The movements of the ancestors are present potentially (768a 14). Under certain circumstances, these potential movements are actualized. Aristotle gives only the most general description of how this works. A relapse usually happens when a certain movement fails to dominate entirely but still does not get mastered (768b26). A male movement can relapse into any of the movements of the ancestors on the male side.

> And if the movements relapse, but the male (*dunamis*) remains, then the [movement]

> of Socrates relapses into the one of his father, and the offspring shall be male and resemble its grandfather or a certain one of the other more remote ancestors. (768a31)

In this case, we have a relapse without being mastered by the opposite. But if we combine the mastery principle with the relapse principle, we get a girl looking like a maternal grandmother:

> If the male (dunamis) gets mastered, there shall be a female and usually it will resemble the mother, but if this same movement also relapses, [it will resemble] the mother's mother or some other remote ancestor according to the same rule. (768a34)

This is how male children get to look like their paternal grandfather and female children like their maternal grandmother. But what about females who look like their paternal grandfather or grandmother and males who look like their maternal grandfather or grandmother?[4]

The answer depends upon how Aristotle conceived the relationship between the mastery principle and the relapse principle. The two explanations quoted above seem to suggest that the mastery principle takes priority over the relapse principle. That is to say, once the dunamis of the father is not mastered (and thus in a sense dominates), then the relapse takes place within the male line on the male side. But if the dunamis of the father qua male is mastered, and the dunamis of the mother qua female dominates, then the relapse takes place within the female line on the side of the mother. But if the mastery principle in this way determines the scope of the relapse principle, Aristotle cannot explain how a boy resembles his paternal grandmother or his maternal grandfather or grandmother. Nor can a girl resemble her maternal grandfather or her paternal grandfather or grandmother. Aristotle must have meant the two principles to operate independently so that the dunamis of any one ancestor (male or female) qua individual could dominate or not. On the level of ancestors, the same rules operate as on the immediate family level. The dunamis of an individual can dominate regardless of the sex of that individual. The same goes for the partial cases of resemblances with ancestors, like having grandfather's crooked nose but nothing else.

> And the same manner is with respect to the parts. For of the parts also some very often resemble the father, others the mother and again others a certain one of the ancestors. For of the parts also of some the movements are present actually and of others potentially, as has been stated often. (768b2)

Like the parts of the parents, so the parts of the grandparents have their dunameis present potentially, ready to be activated.

The last phenomenon mentioned at 769a was the birth of monstrosities. This is simply a case of lots of relapses and very little mastery of any kind.

> For in the end, when the movements relapse, and when the material does not get mastered, what remains is the most general and this is the animal. (769b12)

In that case, we have a monstrosity.

C. Conclusion

It is obvious that in his theory of inheritance, as I have explained it here, Aristotle has qualified his initial form-matter hypothesis greatly.

I think scholars have often failed to take these qualifications into account in their silent and largely negative reactions to the form-matter hypothesis. Charles Singer, for instance, notes that, given this hypothesis, "the inheritance from the mother requires further explanation,"[5] but he fails to inform his reader of the theory of inheritance in Book IV. In the same vein, Mr. Platt has observed that Aristotle's form-matter hypothesis is "fundamentally wrong, for his position is that the mother does not contribute anything resembling the semen of the male, whereas the ovum is just as important as the spermatozoon and carries inheritance equally with it."[6] Again the theory of inheritance in Book IV is ignored. It may well be that Platt is right on Aristotle's being wrong on what the ideal case is, but the depth and seriousness of the mistake is not nearly as large as Platt's remark suggests. Needham also sells Book IV short when he says: "Nor would resemblances between mothers and children suffice to kill this belief [in the dominance of the male] for plants may differ slightly according to the soil in which they are planted."[7] I think that the mastery and relapse principles together with the technical sense of dunamis suggest a far more radical departure from the ideal than a mere adaption to a difference in soil.

Other scholars have gone to the other extreme. Having noticed the qualifications in Book IV, they conclude that Aristotle has overthrown his own form-matter hypothesis. Lesky, for instance, warns that the very discussion of the problems of inheritance is a daring venture for anyone who, like Aristotle, has accepted the form-matter hypothesis.[8] And indeed the expected thing happens. When Aristotle postulates dunameis for each of the ancestors on the mother's side, that clearly, says Lesky, is not any more a case where a passive female reacts to a male stimulus, but the female herself has active dunameis. Lesky believes that at this point Aristotle has broken through his hitherto unbroken form-matter antithesis.[9] I think Lesky has gone too far and has overlooked the function of the relapse and mastery principles. Düring too believes that Aristotle's theory of inheritance, with the potential or latent dunameis of the female and the ancestors, including those on the side of the female, conflicts with his form-matter hypothesis. He bemoans the fact that Aristotle does not seem to have realized that his admissions amount to a rejection of what he had laid down as the generative task of the woman.[10]

In contrast, I do believe that Aristotle knew what he was doing. The ideal is needed to account for the drastic and fundamental differences between the sexes. If he gives us his description of what the ideal case is, he is right back where the pangenesis theory failed when it could not account for these differences. Aristotle's ideal case allows for the fact that the male is generally the active partner (729b13) and, for their differences in reproductive organs, described in chapters one through sixteen of Book I, as well as for the fact that the female and not the male always carries the offspring (716a15). There is a fundamental difference between the sexes and Aristotle's form-matter takes this into account. But he also needs to make the admissions in Book IV, for otherwise his hypothesis flies in the face of what is obviuos to everyone, namely, that not every offspring is a male resembling his father. The qualifications made in Book IV do not conflict with the initial hypothesis, but bring it in line with some facts that would otherwise be its downfall.

This is where the dispute with our Hippocratic author ends. Aristotle finally has met the argument from resemblance. His theory of inheritance explains the detailed cases of resemblance (P) was designed to explain. Not only has Aristotle offered fifteen arguments against (P), but he has also given his own explanation of what initially seems to be very strong evidence on behalf of (P).

FOOTNOTES

[1]Hippocrates, On Ancient Medicine, in the Loeb Classical Library Series, translated by W. M. S. Jones (Cambridge: Harvard University Press, 1923), Vol. I. Joseph Souilhé believes that Aristotle's use of dunamis is fairly close to that of the school of Hippocrates. In that school, the terms had the somewhat technical meaning of "characteristic property." (Étude Sur le Terme Dynamis dans les Dialogues de Platon, Paris: Librarie Felix Alcan, p. 56.) Harold Miller has explored the relationship between dynamis and physis in this Hippocratic treatise in his "Dynamis and Physis in On Ancient Medicine" (Transactions and Proceedings of the American Philological Association, Vol. LXXXIII, 1952). His conclusion is that a dynamis is "a simple, real entity which is characterized and identified by its specific activity and whose specific essential nature is revealed to the senses by its activity. The physis is then composed of an indefinite number of dunameis..." (p. 196). This is indeed quite close to the way Aristotle uses the word in Book IV of the G. of A.

[2]Hans Meyer, "Das Vererbunsproblem bei Aristoteles" (Philologus LXXV, 1919), p. 342.

[3]Erna Lesky, Die Zeugungs-und Vererbungslehren der Antike und ihr Nachwirken (Wiesbaden: Franz Steiner Verlag, 1950), pp. 153/1377.

[4]Lesky thinks Aristotle could not solve this problem and that the Stoics were the first to propose a hypothesis to handle such cases. I argue that Aristotle's theory does allow for a solution if we let the mastery and the relapse principles operate independently from each other. (See Lesky, pp. 154/1378.)

[5]Charles Singer, Greek Biology and Greek Medicine (Oxford: Clarendon Press, 1922), p. 23.

[6]Arthur Platt has translated Aristotle's arguments against pangenesis in A Source Book in Greek Science, edited by Morris R. Cohen (McGraw-Hill Book Company, Inc., 1948), p. 417, note #2.

[7]Joseph Needham, A History of Embryology (New York: Abelard-Schuman, 1959), p. 44.

[8] Lesky, pp. 146/1370.

[9] Ibid., pp. 153/1377.

[10] Ingemar Düring, Aristoteles (Heidelberg: Carl Winter Universitatsverlag, 1966), p. 553.

CHAPTER VI

EVALUATION

A. Preliminary Assessment

The criteria I shall use for evaluating the dialectical dispute before us are those usually recommended for evaluating competing theories. They are: relevance, completeness, unification, evidential support, depth, and simplicity.[1] The first two can be disposed of quickly because on them the antagonists score equally well. The theories and principles proposed on both sides are relevant to the stated problems. If either Aristotle or our Hippocratic author had given irrelevant answers, the other partner would immediately be the victor in the dispute.[2] The same goes for completeness. Given the stated goal of our Hippocratic author, his explanations are just as detailed as those given by Aristotle (assuming that completeness is not the same as depth). Together (P) and (L) cover the ground they do cover in as much detail as the form-matter hypothesis--supplemented by the potential-actual distinction--provides.

The first major difference arises when we measure both theories against the criterion of unification. In On Semen and in On the Nature of the Child, we meet with a somewhat narrow medical and gynecological interest. The author is almost exclusively interested in the origin and development of the human embryo. Though these two treatises have informed large parts of his own treatise, Aristotle has broadened the scope of the discussion to include the entire animal kingdom. And the manner in which he reaches into the furthest and most obscure corners of that kingdom to support his hypothesis is a feat no less brilliant than the one Darwin performed in the Origin of the Species.

What makes the G. of A. one of the first great scientific treatises is not simply the detailed classification of animals according to their mode of reproduction, but far more so the overarching form-matter point of view. In the first sixteen chapters of Book I, Aristotle prepares the way by arguing that the distinction between the sexes is almost universal. In the rest of Book I and in Book II, he proposes, defends, and augments the form-matter hypothesis with respect to man and other oviparous animals. In Book III, he applies these findings to the less perfect, egg-laying animals, and to the bloodless animals. In Book IV, he he tackles problems left over from the argument in Book I. And Book V is more or less an appendix which deals with the secondary characteristics of animals, such as the colour of their eyes, the keenness of their smell, and the curliness of their hair.

Aristotle believed and argued that the form-matter hypothesis applied to the whole animal kingdom. The hypothesis applied quite literally to all those groups of animals that can move and which have males and females. Such animals can reproduce their own kind because they have been generated from the copulation of male and female and hence reproduce themselves in that way. The fixity of the species and, hence, to the existence of a formal (male) and material (female) factor.

> Of the bloodless ones some (i) have male and female; they, therefore, (hōste) generate offspring of their own kind (homogene). Others (ii) generate, but not, however, their own kind. Such are as many as do not come to be from the copulation of animals, but from rotten earth and residues. (715a22-25)[3]

The second group of animals does not have the power of locomotion, does not have the male-female distinction, and does not reproduce its own kind, but first produces a larva out of putrefied matter.

The reason why those who are not themselves generated by the copulation of male and female cannot reproduce their own kind is for Aristotle simply a matter of observation. He has observed no animals which, though they arose by various stages out of rotten earth, were able to reproduce through copulation their own kind. And they cannot reproduce a different kind because that would lead to an endless series of evolving species. "But nature," says Aristotle, "avoids what is infinite. For what is infinite has no end, but nature always seeks an end" (715b15).[4]

There are, therefore, two groups of animals:
1) Those which do have males and females and which, therefore, reproduce their own kind. To this group, the form-matter hypothesis applies strictly; the male contributes the form, and the female, the matter.
2) Those animals which do not have males and females and which, therefore, do not reproduce their own kind. To them, the form-matter hypothesis applies more in spirit

than in letter. Even though they do not have male and female, the form and the matter factors are still present. The form-giving ability that in group 1 exists in the semen of the male is not bound to that specific mode of existence. It can exist independently of any living organism, in the air or in the soil (762b17). In the case of spontaneous generation, out of a mixture of soil and water the heat necessary for concoction might be the heat of the warm season in the environment (762 b15). Apparently, a portion of the soul principle (tēs psychikēs archēs) becomes enclosed and separated in the pneuma, and then forms a fetation out of the putrefied material, planting movement in it (762b17). In this way, Aristotle broadens the scope of the form-matter hypothesis to sedentary animals like barnacles, insects, and eels.

The unification achieved by the form-matter hypothesis is tremendous. Aristotle has brought many hitherto unrelated phenomena together under one theoretical umbrella. He has achieved a unification unequalled by any of his predecessors. If this were the only criterion to be used in evaluating the dialectical dispute between him and our Hippocratic author, Aristotle would certainly have to be judged the victor.

But unification is not everything. If the unifying hypothesis has very little or no evidential support or conflicts with the facts, then, of course, it does not really unify. And if it is bought at the expense of depth and simplicity, then it may not turn out to be a step forward in the progress of science. In the next section, I address myself to the question of evidential support, in sections C and D, to the matter of depth, and in section E, to the matter of simplicity. The conclusion will be that the initial assessment based on the criterion of unification can be sustained.

B. A Bold Conjecture

Throughout this study, I have made occasional references to the philosophical ideas of Karl Popper. This has not been accidental, for I believe that this study of the argumentative structure of the G. of A. supports and is illuminated by Popper's ideas about how it is that scientists proceed and scientific knowledge grows. My analysis also fits such theories as Peirce's[5] theory of abduction and Harman's[6] inference to the best explanation. All of these men have argued that there is no straightforward inductive road from the facts to the theory that is to explain them. They reject the

Baconian kind of empiricism where the facts are collected before the theories are or can be formulated. These authors more or less agree that scientific hypotheses are not simply derived from observation statements (or protocol sentences or sensory reports), but that they are in part a creative product of the human imagination. Says Hempel, "Scientific hypotheses and theories are not derived from the observed facts, but invented in order to account for them."[7]

I think the commentators mentioned in chapter I--those who ignore any possible connection between science and dialectic in Aristotle--have been captured by an inductivist picture of Aristotle's scientific method. This does not exclude a commentator like Richard Mackeon, who believes that, in his methodology, Aristotle maneuvered dialectically between the dogmatism of Plato and the empiricism of Democritus. It turns out that this middle of the road approach nevertheless commits Aristotle to deriving the _archai_ of the various sciences from sensory experience.[8] Like all the other commentators, Mackeon is captivated by the last chapter of the _Posterior Analytics_, to which he refers in support of his interpretation.[9] So, when Mackeon says that Aristotle's scientific methodology is "dialectical," he does not mean, as I do, that the _archai_ of the various sciences are to be derived from the _endoxa_ regarding the problem at hand.[10] The mandate of the _Topics_ is still ignored.

The same inductivist picture has captivated the commentators I mentioned at the end of Chapter II, namely, those who have praised Aristotle as a great scientist but who have been unable to handle his form-matter hypothesis.

This spell can be broken and Aristotle done justice once we realize that it is perfectly all right for a scientific hypothesis to arise out of the creative human mind without being in any direct way derived from the data. What we must look at is the explanatory power a hypothesis has, never mind its origin. It is quite obvious that the form-matter hypothesis was suggested to Aristotle by the production of artifacts. The question is, does this hypothesis have any explanatory power? I think it does.

In fact, if Popper is right, Aristotle's hypothesis has a great deal of this kind of power. Popper believes that the bolder a hypothesis is, the better it is, because it can be tested more easily. By "boldness" he means the specific logical content of the hypothesis.[11] Part of the content of the form-matter hypothesis is, for instance, the proposition that the male supplies none of the material of the offspring. This is a very specific and bold claim and, therefore, easily tested. If it is true, it is part of the truth content of the hypothesis; if it is false, it is part of the falsity content. The better of two competing theories has less falsity content and as much or more truth content than its opponent. In the case of our example, it is a matter of comparing the content of (P) with that of the form-matter hypothesis.

As Aristotle has interpreted (P), it has lots of falsity content. Every one of his fifteen objections in chapter eighteen draws out more of that content. If he can solve the problems solved by (P) and, in addition, solve the ones (P) cannot solve, then his own hypothesis is indeed a bold step forward. Its boldness is connected to its ability to solve the problem of the difference in sex roles, which was one of the main stumbling blocks for (P). It is bold and readily testable because all it takes to be refuted is to show that the male does contribute material to the offspring.

Aristotle is aware of this and systematically sets out to show that his hypothesis is not only testable but that it passes the test and hence has a good deal of explanatory power. He argues that in the case of some insects the female inserts part of herself into the male (723b19). This suggests that the male does not secrete semen from all the parts of his body which he then delivers inside the female. He can do his job without that bulky semen and simply contribute his heat or dunamis (729b28). A little later, Aristotle argues that the phenomenon of wind eggs that were trodden twice shows that the male contributes none of the matter (729b33ff.). If both males had contributed some material part, the offspring would have to have had two such parts, one from each male, but it does not. Hence, the males did not contribute anything material. The same conclusion can be drawn from what happens with oviparous fish (730a18). The milt which the male sprinkles over the eggs laid by the female does not become part of what develops inside the eggs. In all these cases, it is known that the role of the male is needed. Since he does not contribute any of the matter, we may infer that he, therefore, contributes the form of the offspring. In all those cases where the male obviously does secrete semen, that semen functions as no more than a tool for transmitting a movement which at other times may be transmitted without the use of that tool (I, 22). The form-matter hypothesis seems to have passed a severe test implication because, in spite of its boldness, it has not been falsified. It is, therefore, a significant step forward in the progress of science. It seems to have more truth content than (P) and has less falsity content. In short, it has more versimilitude than the theory it replaces.

I do not think it makes a great deal of practical difference whether we describe what Aristotle does as verification or falsification. Theoretically, it may make a difference since, as Hempel points out, the logic of falsification involves a legitimate modus tollens: "If H is true, then so is I. But (as the evidence shows) I is not true. Therefore, H is not true."[12] The logic of verification involves, however, the fallacy of affirming the consequent. "If H is true, then so is I. But (as the evidence shows) I is true. Therefore, H is true."[13] Theoretically, therefore, it is better to say that Aristotle attempted to falsify his hypothesis and failed rather than saying that he attempted to verify it and succeeded. The former takes the form of explaining away the apparent counter-examples of the cases where the male contributes bulk-like semen. The bulk plays no role and is not a necessary ingredient. In fact, it evaporates (737a15). The latter cannot succeed because the problem of induction prevents us from ever totally verifying a hypothesis. Further investigation may always show it to be false after all. The language of falsification is preferable because it takes the problem of induction seriously and "solves" it by admitting it.[14] Given the severe testing of a theory, however, it makes very little practical difference how we describe the procedure. It seems quite natural to say that a series of failures to falsify add up to a success at verifying. We may say, therefore, that Aristotle's form-matter hypothesis has been verified because, though he tried, he failed to falsify it.

Furthermore, the form-matter hypothesis has more versimilitude than (P) even if it too, just like (P), should later turn out to be false after all, as it indeed has. In spite of the curious fact that the sperm cell of the male is tightly packed with heriditary material and the egg cell contains the food and nourishment for them both after their union,[15] there is no evidence today to support the natural and ideal dominance of the male. It is quite evident that both the sperm cell and the egg cell contain their own heriditary material and that the one in no way naturally lords it over the other. This does not totally disqualify Aristotle's inference from the difference in sex organs to a difference in function. The inference may or may not be correct. On the one hand, one and the same function may be performed in more than one way or with more than one kind of tool. On the other hand, not until the microscopic processes of molecular biology are shown to underlie the development of the macroscopic phenomena Aristotle was worried about, can we tell for sure whether the inference, which in effect is an appeal to the notion of final cause, is warranted. I return to this point in section D where I deal with the charge of vitalism.

If any of the ancient authors was right, Empedocles comes closest to the truth. For he believed that the male and the female each contributed half of the (heriditary) material. This perfectly fits the fact that egg cells and sperm cells contain only half sets of chromosomes, which sets are fused in the fertilized ovum. Aristotle simply failed to imagine the microscopic processes that might make Empedocles' theory work. And even if he had imagined it, there would have been no way in his day to test such a conjecture.

But none of this contemporary input changes my claim that Aristotle's hypothesis was a step forward in the history and progress of science. Popper's schema of the growth of scientific knowledge can be of some help here:

$$P_1 \rightarrow TT \rightarrow EE \rightarrow P_2$$ [16]

In our case, P stands for the general problem of what the male and female roles are in reproduction. This problem was widely discussed in antiquity.[17] Why males and females exist and are both needed to reproduce was pretty much an open question in Aristotle's day, so he proposed a tentative theory (TT) to the effect that the male supplies the form and the female the matter of the offspring. Having proposed the theory, he sets out to eliminate the possible errors in it (EE). This includes an evaluation of the main competing theory according to which the sexes play similar roles in the reproductive process. Aristotle refutes that theory and shows how his own hypothesis solves the problem of the sex roles and passes the required test implication. After this, more problems arise, such as the questions of development and life (P_2). And again new theories are tentatively proposed and, hopefully, shown to be successful. My point is that, given the background information against which Aristotle wrote, his form-matter hypothesis was a step forward. Temporarily, at least, its boldness and, therefore, its explanatory power was upheld and sustained.

Testability is, of course, not the only requirement for being the better of two theories. An even more basic (because prior) requirement is that the theory solve the problems its competitor had already solved. Not much progress has been made if the form-matter hypothesis solves the problem of the sex roles but fails to solve the problem of detailed resemblances between the offspring and the mother, to which (P) had been addressed. The same goes for the problem of development and the apparent success of (L). It does not help if one significant problem is solved at the expense of relegating several other ones back to the unresolved status.

I discussed Aristotle's answers to these other problems in Chapters IV and V. It is quite evident that Aristotle supplements the initial form-matter hypothesis with quite a few other distinctions and maxims. At a juncture such as this, Hempel sounds a note of warning. He feels that the more additional entities and maxims need to be added to the original hypothesis the closer it comes to being *ad hoc*.[18] Though there is no absolute test for sorting out *ad hoc* hypotheses from others, if a hypothesis needs to be bolstered by too many qualifications and additions, its viability is threatened. This warning can be translated into the terms of our example with the following three questions. 1) Granted that the form-matter hypothesis is initially a viable hypothesis, does not the need to supplement it with the potential-actual distinction undercut this viability? Are not all explanations based on this distinction vacuous and does not this vacuity rub off on the initial hypothesis? 2) Does not Aristotle's appeal to *psychē* and *pneuma* open him up to the charge of being a vitalist? How can any hypothesis that involves such mysterious things as vital forces be a step forward in the progress of science? 3) Is not Aristotle's appeal to *dunameis* in Book IV one grand scheme to save his form-matter hypothesis from what otherwise would obviously be adverse evidence? Is the salvation worth the price of these wild and unconfirmed speculations?

In the next three sections, I discuss these three questions. My conclusion will be that the additional machinery does not undercut but adds to the initial plausibility of the hypothesis and thus underscores the preliminary assessment of section A.

C. The Vacuity Charge

The argument containing the charge goes something like this. When we are told that opium puts us to sleep because it has a certain power (*virtus dormitiva*) to do so, we are given no explanation at all. An appeal to such an occult and mysterious power is nothing but an admission of ignorance. Anything whatever can be explained in this way. All we need to do is to claim that the thing had the power to do whatever it is we want explained. No one will ever be ignorant and we will always have explanations for everything in a universe filled with mysterious and occult powers. No, unless the *virtus dormitiva* of opium can be identified and so described independently of its alleged effects, the explanation of why it puts us to sleep is nothing but an *ad hoc* escape. Without that additional identification of the sleep-inducing power, the explanation not only lacks depth; it is not even an explanation.

And there seems to be very little (if any) difference between this *virtus dormitiva* of opium and the *dunameis* or powers to which Aristotle appeals in his explanations. He has invented these active and passive powers simply to cover up his ignorance. This is shown by the fact that he makes no attempt to identify and describe these powers independently of their alleged effects. His explanation is, therefore, just as vacuous as the one of the opium example; while explaining nothing, he has revealed to us his ignorance.

The charge hurts in more ways than one. There is the philosophical charge that Aristotle's explanation is vacuous. Something must be said in response to this charge. But there is also the historical charge. The vacuity charge leads us to the heart of an ongoing discussion in the philosophy of science about explanatory uses of dispositional properties. Such properties are meant to explain their displaying episodes but, so the argument goes, do not do so unless independently identifiable. With some liberty of interpretation, it might be claimed that when Aristotle first introduced the potential-actual distinction in his explanations, he, in fact, introduced what are now called dispositional properties. After all, there seems to be very little difference between the present day distinction of dispositional properties and their displaying episodes, and Aristotle's distinction between *dunameis* and their actualizations or activities. Aristotle especially is, therefore, to be blamed for having polluted the language of philosophers and scientists for many centuries. It was not, the attackers might add, until Nelson Goodman declared in the mid-fifties "the passing of the possible"[19] that we could finally throw off the burden of possibles imposed upon us by Aristotle. Aristotle's philosophical mistake had, so it seems, grave and long-lasting consequences.

Without pretending to settle all or most of the issues raised in the debate about dispositional properties, I will defend Aristotle's appeal to and use of the potential-actual distinction. I shall argue that the vacuity charge as stated above is misplaced when leveled against Aristotle. I begin with a contemporary statement which comes close to what I take Aristotle to be saying in his talk about powers.

In a paper entitled "Powers," Harré has argued "that not only are powers indispensable in the epistemology of science, but are the very heart and key to the best metaphysics for science."[20] Harré believes that when one thing has the power to behave in a certain way, and another thing does not, that these two things differ not just in the way they behave, but also in what they are in their intrinsic nature. He unpacks the ascription of a power as follows:

> X has the power to A = if X is subject to stimuli or conditions of an appropriate kind, then X will do A <u>in virtue of its intrinsic nature</u>.[21]
>
> The satisfaction of intrinsic enabling conditions ensures that a thing or material is of the right nature for the exercise of a certain power.[22]

All of this sounds very much like what Aristotle tells us in Book IX of the *Metaphysics*. There he says (chapter 7) that when the principle of movement (*archē kinēseōs*) of a thing (say X) is external to X, then X is potentially (*dunamei*) Y, if nothing internal to X prevents X from becoming Y. In this case, the power in X is a passive power. On the other hand, if the principle of movement is internal to X, then X is potentially Y if nothing external to X prevents X from becoming Y. Here X has an active power. The distinction drawn here between internal and external factors in generation and change is very much like the one Harré draws between external stimuli and internal enabling conditions.

The point of ascribing a power to something is, Harré believes, "to make the intrinsic nature of the things and materials involved...in more or less measure responsible for that phenomenon occurring."[23] This is what I think Aristotle had in mind in his repeated objections to Empedocles. He blamed Empedocles for using Love and Strife as explanatory powers and for not making a sufficient study of the very nature (ousia) of the elements themselves.[24] Had he done so, Empedocles would have discovered, as did Aristotle, the intrinsic enabling conditions of the elements and the need for Love and Strife would have been obviated. Aristotle does not object to the explanatory use of powers, but he wants them linked to the nature of the things to be explained. In Physics II, 1, he links the concepts of physis (nature) and dunamis (power) quite closely.[25] Aristotle generally does with powers what Harré says their function is, namely, to draw attention to the nature of the things to be explained.

But Aristotle and Harré part company on the kind of attention so focused. For Harré, the attention comes in the form of a promise. In the case of the opium example, for instance, the dispositional statement is a promise "which says that the cause of sleepiness is not to be put down just to the fact of ingestion of opium, but is to be looked for in the nature of opium itself...."[26] Power talk is promise talk about the nature of the thing which is said to have this power.

> To ascribe a power is to assert a specific behaviour hypothetical together with an unspecified categorical referring to the nature or constitution of the thing or material concerned.[27]

He adds that this "categorial component is like a promissory note; we need only believe that it is not logically impossible for it to be cashed."[28] What we are promised in the ascription of a power is that sooner or later we shall be able to explain the possession of that power in terms of a categorial or nondispositional property of the thing having the power. In this way, the vacuity charge is met, for an appeal to powers does not pretend to be a full-fledged explanation. It is only the promise and schema of such. Says Mackie, in the same context of answering the vacuity charge: "There is nothing absurd or improper in postulating this [sleep inducing power], and (as Harré says) this is just how scientists proceed; such a postulation leads to further discoveries."[29] Though not an explanation in itself the postulation of a power points the way to a genuine explanation. For Mackie, too, the vacuity charge is met by admitting that an appeal to powers does not constitute a real explanation, but only the schema or promise of one.

At this point, it would seem that Aristotle is in trouble because he did not see his own power talk as promise talk. There is every indication that he took his equivalences of our dispositional statements as providing real explanations. When he appeals to the potential-actual distinction to explain how an animal develops from the female matter, he thinks he has already given an explanation and not merely a schema to be fleshed out by further research. He had already done the research and decided that the bloodlike character of the menstrual blood and the pneuma contained in the male semen together explained the phenomenon of development. There is no hint of tentativeness in his argumentation. Now, since the vacuity charge is met with a promissory note, and since Aristotle did not give such notes, he cannot be said to have met the charge of vacuity. A man cannot be said to have made a promise unless he is conscious of having done so.

It is tempting to get Aristotle out of trouble by ascribing the finality and lack of tentativeness of his power talk to his exuberance and enthusiasm for his work. We might say on his behalf that, even though he did not think of his explanations as merely pointing the way, this is, in effect, what he did do. It often happens in the history of science that what one thinker thinks is pretty much the final answer in retrospect turns out to have been no more than a crude approximation and pointer in the right direction. Think, for instance, of how the Mendelian gene has turned out to be a segment of DNA.[30] So rather than say that Aristotle's explanations were vacuous, we should say that, in spite of his own misplaced finality, he succeeded in pointing the way to further and fuller explanations of later days and centuries. This is a tempting defense to mount and there is a lot of truth in it, but it does not do enough justice to Aristotle's scientific acumen. It blames him for having given final answers where he should have given promissory notes instead.

I resist this way of defending Aristotle because I think that the vacuity charge as stated above does not apply to him, and that there was no way he could have given promissory notes, even if he had wanted to do so.

The charge, as discussed by Harré and Mackie, supposes that scientists have already discovered some sort of deeper, let us say molecular, structure of at least some things. It is against this background knowledge that the suspicion of powers and dispositional statements arises. Nelson Goodman can call dispositions "occult" and "mysterious" because he lives in the twentieth century.[31] The vacuity charge supposes that such a deeper level has or can be reached, which is precisely why contemporary philosophers of science present power talk as promise talk. The promise is extracted against the background knowledge that opium has a molecular structure which really is the cause of its sleep-inducing capacity. Against such background knowledge, the charge of vacuity, when powers are offered as final explanatory entities, makes sense. If there are such things as molecular structures, which are known to be responsible for the behaviour of things, then there is no need to be satisfied with the ascription of a mere power. Even this does not mean that there are no powers. It means that on this particular level, powers have been superseded as the fundamental explanatory entities in science.

It seems to me that in terms of Carnap's[32] division between internal and external questions, the vacuity charge raises an external but not an internal question. Using his example of thing language, Carnap believes that it makes perfectly good sense to ask whether there are such things as dinosaurs. But given that language, it makes no cognitive sense to ask whether there are such things as things, for the language has stipulated that there shall be things. This last question is an external one and raises the issue of how we are to evaluate competing a priori or stipulative schemes. Carnap says that "to accept the thing world means nothing more than to accept a certain form of language, in other words, to accept rules for forming statements and for testing, accepting or rejecting them."[33] But then he adds that the decision to accept the thing language itself depends on "the efficiency, fruitfulness, and simplicity of the thing language."[34] To answer an external question depends thus on the utility of the scheme within which the objectionable phrase or word occurs. Though the acceptance of the scheme involves matching it against the facts, there are no established procedures for doing so. Efficiency, fruitfulness, and simplicity are all very vague and broad criteria for answering external questions. Nevertheless, they are of help in defending Aristotle against the vacuity charge.

To level the vacuity charge against Aristotle is to raise an external and not an internal question, which is why I did not want to take the easy way out previously. There were, in his day, no established procedures for translating power talk into talk about other even more fundamental entities. It is not as if Aristotle lived in a day and age where scientists had already penetrated to deeper levels of reality as they have done in our age. It is not as if Aristotle should have been suspect of powers but was not. And it is not that he should have regarded all his power talk as promise talk but did not. None of these allegations apply in the absence of the kind of scientific progress with which we are familiar. Words like "occult" and "mysterious" are relative to the progress of science. What in one age is occult may well be a significant step forward in another age. The way Goodman and others look at powers was roughly the way Aristotle looked at Empedocles' Love and Strife. Those were for him the occult powers to be banished from the practice of science. Aristotle appealed instead to the powers of things to be explained, avoiding the supernatural. At that time, power talk was the deepest level to which scientists had penetrated. There was, therefore, no background knowledge against which to raise the vacuity charge as an internal question. No power had yet been reduced to or explained in terms of some other deeper level. There were the observed macroscopic phenomena and the powers with nothing between or below.

We are left, then, with the question of whether in Aristotle's day power talk was fruitful, efficient, and simple, and on this score, there need be no doubt. All of these criteria add up to the question of utility. Clearly, Aristotle's appeal to active and passive powers was extremely useful for the progress and practice of science. This talk served to draw attention to the need for explanations in terms of the phenomena of nature themselves. In a day when science was still in its infancy, and when many myths still had to be cast out, Aristotle's potential-actual distinction was a definite step forward because it was a step into the nature of things. We now know a lot more about the nature of things and, consequently, do not postulate our powers as closely to the macroscopic level as Aristotle did. But chances are that powers and dispositional properties are still postulated and are still appealed to as explanatory entities on what today are the frontiers of knowledge. The question is not whether Aristotle was right or wrong in believing in the existence of powers, but whether he postulated them in the right place. And here we must answer yes, because he placed them where every scientist places them, namely, at the frontiers of knowledge. The question of whether powers or dispositional properties exist is and always has been an external question. And such talk will always be useful and even necessary as long as we do not know everything there is to know about the natures of things.

D. The Charge of Vitalism

Hans Driesch, who more than anyone gave Aristotle the reputation of being a vitalist, succinctly stated the question of vitalism as the question of whether the purposive character of life processes "is the result of a special constellation of factors known already to the sciences of the inorganic, or [whether] it is the result of an autonomy peculiar to these processes themselves."[35] It is the question of the autonomy of biology. Vitalists believe that there are at least some laws or principles or entities in biology which cannot be reduced to or replaced by those of chemistry and physics. Mechanists or materialists, on the other hand, believe that such reduction, although not yet fully accomplished, is a distinct possibility and should certainly be accepted as a heuristic maxim in research.

Most scientists and philosophers today do not believe in the autonomy of biology. They probably agree with Hull's remark that life is like magnetism in that "both result from the organization of the material out of which they are made."[36] The views expressed by Simmias in the Phaedo are back in vogue today.

The objection to vitalism is not that vital forces, entelechies, or the soul-principle cannot be seen, heard, or felt, for the same is true of electrons and a host of other theoretical entities. It is that the theory lacks predictive power. Says Hempel:

> [The doctrine of vitalism] does not indicate under what circumstances entelechies will go into action, and specifically, in what way they will direct biological processes: no particular aspect of embryonic development, for example, can be inferred from the doctrine, nor does it enable us to predict what biological responses will occur under specified experimental conditions. Hence when a new striking

> type of 'organic directiveness' is encountered all that the neovitalist doctrine enables us to do is to make the *post factum* pronouncement: 'There is another manifestation of vital forces!'; it offers no grounds for saying: 'On the basis of the theoretical assumptions, this is just what was to be expected--the theory explains it!'[37]

This charge depends, of course, on the symmetry between explanatory and predictive power. This symmetry thesis is not beyond debate.[38] It seems to me that reasons explain human actions perfectly well but they cannot be used to predict those same actions in the manner required by the covering law model. This at least opens the door to the possibility of there being explanations in biology which do not have the customary predictive power. Still, there can be no doubt that the doctrine of vitalism does not yield any good research projects. Since, as Hempel says, all its explanations are *post factum* pronouncements, it can hardly function as a medium of progress. As a heuristic maxim, any scientist worth his or her salt will accept reductionism. The question is whether in this respect Aristotle was worth his salt in the Fifth Century B. C. I think he was, even though he did not accept the heuristic maxim of reductionism. I shall argue my case largely in terms of the argument presented by Jacques Monod in his book *Chance and Necessity*.

The main reason why Monod's argument is relevant to an evaluation of Aristotle's position is that he does not belittle, ignore, or seek to translate away the purposive character and function of living beings and their parts. All the functional talk and final causes of the *Parts of Animals* are perfectly compatible with what Monod has to say. He thinks that living beings are "objects endowed with a purpose or project"[40] and refers to this fact as the teleonomy of living beings.

A second striking feature of living beings is that their shape and structure is the result of an internal autonomous force and is not caused by any external force as is the case with artifacts. This is also the way Aristotle explains the difference between natural organisms and artefacts.[41] Monod calls this the autonomous morphogenesis of living beings.

The third characteristic of living beings is "their ability to reproduce and to transmit *ne varietur* the information corresponding to their own structure."[42] Aristotle was also impressed by this fact. Indeed, he was so impressed by the phenomenon of invariance that he thought the species to be eternal (G. of A. II, 713b18 ff.). These three features (teleonomy, autonomous morphogenesis, and invariance) together constitute what makes living beings unique, and all of them play a large role in Aristotle's biological deliberations.

Vitalism clashes with modern science in the following way; the latter operates with what Monod calls "the postulate of objectivity," which is nothing but a dogmatic unprovable[43] assertion that final causes have no real causal efficacy. Nevertheless, the scientist cannot help but note the teleonomic activities and functions of living beings. He is, therefore, in a bind; some of the most obvious data conflict with his most fundamental postulate. He escapes by explaining teleonomy as caused by and due to invariance. The vitalist, on the other hand, operates on the reverse hypothesis "that invariance is safeguarded, ontogeny guided and evolution oriented by an initial teleonomic principle, of which all these phenomena are purported manifestations."[44] The issue is joined on whether teleonomy precedes invariance or invariance teleonomy.

In chapters V and VI, Monod comes to the conclusion, stated much earlier, that "the mechanism [of invariance] is sufficiently well known today for us to be able to state that no nonphysical principle is required for its interpretation."[45] This, if true, means that invariance precedes teleonomy and that the latter cannot be the cause of fixity (or eternality) of the species. The same microscopic processes that explain invariance can also, in principle, explain teleonomy to us. Says Monod, "All these teleonomic performances rest, in the final analysis, upon the proteins' so-called 'stereospecific' properties, that is to say upon their ability to 'recognize' other molecules (including other proteins) by their shape, this shape being determined by their molecular structure."[46] There is a point to this, because invariance itself is a prime example of teleonomic activity. The less ignorant we are of these microscopic processes, the less room there will be for the speculations of the vitalist.[47] The implication of Monod's argument is, however, that not until these microscopic processes were discovered and invariance linked to the structure of DNA could we truthfully say that invariance precedes teleonomy. Now, however, having discovered no entelechies, vital forces, or soul principles and yet having explained invariance mechanically, we can say that invariance and not teleonomy is the more basic of the two properties in living beings. Since vitalism thrives on ignorance, it must yield to the advance of scientific knowledge.

I agree with this line of attack, but want to stress the other side of the coin. This is that in the absence of knowledge of these microscopic processes of molecular biology, vitalism is a viable theory because it seeks to account for the obvious teleonomic character of living beings and their parts. And this is precisely the position of Aristotle. In the absence of this new knowledge, it makes perfectly good scientific sense to suppose that teleonomy precedes invariance. Driesch did well to connect the motivation for vitalism with the observance of purpose in nature. This character of living beings is what led Aristotle to introduce the various souls (especially the vegetative and sensitive ones) as explanatory factors in the development of a plant or an animal. Heating and cooling, which for him were the most basic mechanical processes, could, he thought, be the cause of hardness, softness, and brittleness, "but they certainly cannot produce the logos that makes one thing flesh and another thing bone" (G. of A. II, 1, 734b33). This logos, form, or formula is contributed by the movement of the male parent which brings with it the sensitive soul (734b35 and 741a12). It is the sensitive soul principle contributed by the male which guides the development of a living, unified animal, which "sees to it" that hands, a face, and feet develop at the proper time and in the proper place.

Our Hippocratic author was also impressed by the phenomenon of invariance and he reverted to the like-naturally-goes-to-like maxim as the underlying mechanism that caused the phenomenon. But, as Aristotle pointed out, he could not explain what made each portion of fluid go to its appointed place. Empedocles made a similar mistake when he assigned to Love the power to unite and to Strife the power to separate the different elements. In this way, he left the cause of the exact mixture unexplained. In both these systems, as well as in the atomism of Democritus, chance and not lawlike necessity ends up being the fundamental cause of invariance. In retrospect, it turns out that chance does play a crucial role in mutations and perturbations and hence, eventually, in the development of new kinds of invariance. Because of the accidental alterations that can take place in a DNA sequence, chance is a major factor in the microscopic processes underlying invariance. On the other hand, the teleonomic structures lay down the initial survival chances of any accidental alterations, for such alterations must not interrupt the coherence of the teleonomic system.[48] Furthermore, "once incorporated in the DNA structure, the accident--essentially unpredictable because always singular--will be mechanically and faithfully replicated and translated...."[49] There is an interplay here between chance and necessity. And given the element of chance, not evolution, the stability of the species is the real paradox and source of amazement.[50] While his opponents unwittingly left the door open to chance and to a certain extent were correct in doing so, there can be no doubt that Aristotle was justified in pressing on toward a clarification of the element of necessity. The teleonomic activities and functions of living beings are so obvious and so clearly not simply or fundamentally due to chance only that Aristotle has every methodological right to search for a better explanation, even if it meant drawing an analogy between art and nature. For in the absence of the knowledge of the necessity contributed by DNA to the phenomenon of invariance, the soul-principles invoked by Aristotle are not all that far fetched. Aristotle sought to explain the lawlike character of nature and could find no cause for it in the mechanical processes known to him. Just like it is hard to see how mere heating and cooling can make a sword, it is hard to see how these processes can make a hand or a face. The only difference is that in the case of artefacts, the form-giving movement lies outside the product, while in nature it lies inside the thing produced (G. of A., II, 1, 735a).

Then, too, it may be said on behalf of Aristotle, that if vitalism thrives on ignorance, there is still plenty of food left even today. Monod admits that "while embryologists have provided admirable descriptions of development, [they] are a long way yet from knowing how to analyze the ontogenesis of macroscopic structure in terms of microscopic interactions."[51] Though a lot is known about some microscopic processes, these processes have not yet been linked up to the development of tissues and limbs.[52] Yet those macroscopic characteristics, uniform as well as nonuniform (to use Aristotle's terminology) are precisely the aspects of teleonomy that interested Aristotle most. To charge, as Hempel does, that vitalism fails because it lacks the power to predict, loses a lot of its sting once we realize that those telenomic activities and functions about which Aristotle worried still have not been explained in terms of the covering law model. It is not the case that the vitalist explanation has to compete with another more adequate explanation. We cannot definitely say that Aristotle was wrong in giving fundamental explanations of the phenotypic traits of living beings and of backing them up with soul-principles as their efficient causal agents.

The case in biology is much like the example Davidson gives of a hurricane that caused a catastrophe. Says Davidson:

> The laws needed to predict the catastrophe with precision would, of course, have no use for concepts like hurricane and catastrophe. The trouble with predicting the weather is that the descriptions under which events interest us--'cool, cloudy day, with rain in the afternoon'--have only remote connections with the concepts employed in the more precise known laws.[53]

Analogously, the descriptions under which Aristotle was interested in the teleonomic activities of animals--eye, hand, heart, foot, etc.--are, by the very admission of even the most ardent reductionists, only remotely connected with the descriptions of the presently known microscopic processes of molecular biology.

E. Dunameis as Theoretical Entities

The third and last question at the end of Section B concerned the theory of inheritance in Book IV. I briefly raised the possibility that this theory is no more than an extravagant attempt on the part of Aristotle to save an otherwise doomed hypothesis. Did Aristotle try to do by the force of fancy in Book IV what he could not do by the force of argument in Book I? In the present section, I respond to this charge and argue that what Aristotle does in Book IV is perfectly in keeping with accepted scientific practice.

I purposely reported Aristotle's views as adding up to a <u>theory</u> of inheritance. Carl Hempel has described the purpose and function of a theory as follows:

> Theories then seek to explain those [empirical] regularities and, generally, to afford a deeper and more accurate understanding of the phenomena in

> question. To this end, a theory construes those phenomena as manifestations of entities and processes that lie behind or beneath them, as it were. These are assumed to be governed by characteristic theoretical laws, or theoretical principles, by means of which the theory then explains the empirical uniformities that have been previously discovered and usually also predicts 'new' regularities of similar kinds.[54]

It seems to me that Aristotle's discussion, as reported in the preceding chapter, does exactly what Hempel says a scientific theory should do. The dunameis, which Aristotle has introduced, function as theoretical entities and the three axioms (potential-actual, mastery, and relapse) are the microscopic processes that govern the behavior and interrelations of these microscopic entities. Aristotle has never seen any of these entities or directly observed any of these processes, but he has postulated them to be able to account for the phenomena of resemblance he did observe. He has construed the empirical phenomena "as manifestations of entities and processes that lie beneath them, as it were." I think we have here one of the earliest recognitions of the difference between phenotype and genotype levels, with the usual difficulties attached to that distinction.

My point, however, is that Aristotle's difficulties are not unusual and cannot be discarded as merely an attempt to save the form-matter hypothesis. Of course, he is trying to save that hypothesis, but the means by which he does it are honorable ones that fall within the pale of normal scientific practice. Aristotle essentially did what Mendelians[55] did when they postulated the gene as the basic, microscopic hereditary factor and went on to speak of how such genes can be dominant or recessive. All of this was done to explain the observed cases of detailed resemblances.

Darwin's theory of pangenesis provides another example. In fact, there is very little difference between this theory and the one adopted by Aristotle. Though Aristotle attacked pangenesis and sought to replace it in Book I with his own form-matter hypothesis, in Book IV he comes up with a modified pangenesis theory within the framework of his earlier hypothesis. Darwin felt that he could explain a lot of otherwise puzzling phenomena with what he called "the provisional hypothesis of pangenesis."[56] According to this hypothesis:

> Each unit of each [bodily] part, as it changes during development...must throw off gemmules. All organic beings, moreover, include many dormant gemmules derived from their grandparents and more remote progenitors, but not from all their progenitors. These almost infinite numerous and minute gemmules must be included in each bud, ovule, spermatozoon, and pollen grain....each separate cell [of the body] generates its kind.[57]

A gemmule thrown off during a certain period in the life of the parent will develop, if at all, during a similar period in the life of the offspring. This happens because certain gemmules have a natural affinity for those others whose cells were proximate in time and place to the cells from which they were thrown off. Some gemmules are dominant, and others dominated. The more gemmules from a certain parent or ancestor are dominant, the more the offspring will resemble that person. All of this sounds very much like Aristotle's <u>dunameis</u> and the way they master, are mastered, or relapse.

Aristotle's theory does not predict any future cases of resemblance. It fails to do this, mostly because of a lack of adequate bridge principles between the microscopic entities and processes and the observed phenotypic traits. But the same thing is true of Darwin's theory of pangenesis (and, for that matter, to a large extent of his theory of evolution) and also of Mendelian genetics. In the former, the dominance and latency of a gemmule has to be inferred from the presence or absence of a phenotypic trait. The processes by which the gemmules are said to interact are not sufficiently connected to the phenomena to be explained for us to be able to predict the latter on the basis of the former. The same trouble has plagued Mendelian genetics. The gene was construed as a theoretical entity inferred from the phenotypic trait.[58] It is true that statistically Mendel's experiments with peas allow us to make some rough guesses, but these guesses are not based on a deeper understanding of the processes of dominance and recession by which genes are supposed to interact. The necessary bridge principles are still missing.

Aristotle is not alone in his failure to connect the genotype processes to the phenotypic traits in such a way that the latter can be predicted on the basis of the former. As I mentioned in the preceding section, scientists cannot make these kinds of predictions even today. It is true that Aristotle's theory of inheritance lacks a certain predictive power. But so does every other theory that concerns itself with phenotypic traits. No doubt enormous progress has been made in understanding what happens on the genotype level. We have moved from *dunameis* to gemmules to genes, and now to segments of DNA. Each step was a step deeper into the structure of the microscopic world responsible for the macroscopic properties we seek to explain. We cannot fault Aristotle for not having bridged the gap between these two worlds. Rather, we should commend him for having noticed that gap and for having drawn attention to it.

This is where Aristotle improved upon the Hippocratic author. Though Aristotle borrowed much from him, he also improved much upon what he borrowed. The pangenesis theory as presented in *On Semen* cannot account for the detailed cases of resemblance between offspring and long-gone ancestors, because the components of the semen (whether they be parts of fluid or parts of the body) are understood and wholly defined in terms of what is contributed just by the bodies of the two parents. No provision is made for wider connections. Against this background, Aristotle's new conception of *dunameis*, as what we would call theoretical entities, is a distinct improvement. Aristotle recognized a gap between the genotypic and phenotypic levels of reality. Though we have made much progress in bridging that gap, some important links are still missing. And without those links, no theory in biology or genetics will ever be able to meet the symmetry thesis implicit in the covering law model. As in the days of Aristotle, most resemblances are still being explained without being predicted.

F. Conclusion

I conclude that the preliminary assessment based on the criterion of unification still stands. Aristotle's form-matter hypothesis was a bold step in the history of science. It was a bold step because his belief that the male provides none of the material could, if false, easily be shown to be so. Aristotle was, however, able to point out that in the cases of wind eggs that are trodden twice and, in the case of oviparous fish, males play a crucial role, but provide none of the material. Hence, when they do supply bulk-like semen, that semen serves as a mere tool for the transmission of the form-giving movement. It was a bold step because Aristotle drew a clear line of demarcation between the contribution of the male and the one of the female.

It was a step forward because in its very boldness lies Aristotle's answer to the problem of sex differentiation which had puzzled many ancient thinkers, including our Hippocratic author. No satisfactory and realistic solutions had been offered to the problem of why a male and a female are needed in procreation and the problem of what the difference is between them. Aristotle's hypothesis solves this problem. And as such it is a distinct improvement over the pangenesis theory. If the semen is made up of parts from all the parts of an animal, there is no way of explaining why the female needs the male and the male needs the female in order to procreate. Most of this chapter was devoted to arguing that the bold conjecture of Book I is not undercut or rendered meaningless by the qualifications Aristotle makes in Books II and IV.

The form-matter hypothesis does not solve all the problems by itself. It does not explain how the male imposes his form on the female matter without being in continuous contact with that matter. To solve this problem, Aristotle introduced the potential-actual distinction. The male, who is actually like the offspring, by way of a movement or power, transmits his form to the material of the female, who is potentially the offspring. In section C above, I argued that his use of the potential-actual distinction does not open Aristotle up to the charge of having given a vacuous explanation. Rather, his talk about both active and passive powers of things had a demythologizing effect on the history of science. With it, Aristotle drew attention to the fact that explanations of natural phenomena must be given in terms of those phenomena themselves.

Contrary to what we expect today, Aristotle's vitalism does not conflict with this naturalistic emphasis. Aristotle's pneuma and its soul-principles were meant to explain the fixity of the species. While Aristotle's predecessors and contemporaries left too much to chance, Aristotle was well within the bounds of normal scientific practice to press on for an explanation of necessity, order, and purpose. In the absence of present day advances in molecular genetics, it was not a mistake for him to suppose that teleonomy precedes invariance.

Finally, Aristotle's theory of inheritance is not merely an ad hoc scheme to keep the form-matter hypothesis from flying in the face of the obvious facts. Though it does save the theory and does draw a distinction between the ideal (a boy looking exactly like his father) and the real (all other cases of resemblance), this theory is not an ad hoc device. For one thing, the distinction between the ideal and the real is made plausible because it does explain the differentiation of the sexes. Secondly, the theory is an early and sophisticated attempt to explain phenotypic resemblances in terms of genotypic processes.

Since the different charges of vacuity, vitalism, and ad hoc qualifications have been met, Aristotle's explanation of the generation and development of animals must be judged superior to the one offered by our Hippocratic author.

FOOTNOTES

[1]These happen to be borrowed from Peter Achinstein's paper "Explanation," section IV, reprinted in Studies in the Philosophy of Science, edited by Nicholas Rescher (Oxford: Basil Blackwell, 1969), chapter I.

[2]For example, if my initial interpretation of the argument in Book I as being eristic had prevailed, then Aristotle could have been charged with giving an irrelevant answer. See note 45 of chapter II.

[3]Aristotle did not think there were exceptions to the female-male distinction among the blooded animals (III, 5, 741a33-38).

[4]This is, in effect, an argument against the evolving of one species out of another. See also H. B. Torrey's and Frances Felin's paper "Was Aristotle an Evolutionist?" in The Quarterly Review of Biology, March 1937.

[5]Charles S. Peirce, Essays in the Philosophy of Science, Chapter XIII, "The Logic of Abduction." (Indianapolis: The Bobbs-Merrill Co., Inc., 1967). "I am quite sure that you must be brought to acknowledge that man's mind has a natural adaptation to imagining correct theories of some kind..." (p. 238).

[6]Gilbert H. Harman, "Inference to the Best Explanation," Philosophical Review, January, 1965. Harman argues that induction really is a case of an inference to the best explanation: "One is warranted in making this inference [from all observed A's are B's, to all A's are B's] whenever the hypothesis that all A's are B's is [in light of all the evidence] a better, simpler, more plausible [and so forth] hypothesis than...[other hypotheses]" p. 90).

[7]Carl G. Hempel, Philosophy of Natural Science (Englewood Cliffs: Prentice-Hall, Inc., 1966), p. 15.

[8]Richard Mackeon, "Aristotle's Conception of the Development and the Nature of Scientific Method," Journal of the History of Ideas, January 1947. "The dialectical opposition of the doctrines of Democritus to the dialectic of Plato is, consequently, a convenient device to characterize his own doctrines as a mean between the two extremes, precisely because it indicates scientific as well as dialectical problems and the

scientific statement of Aristotle's own position, as justified not by opposition to what men say but by the evidence of the facts..." (pp. 23-24).

[9] In the Posterior Analytics, "The path from sense-perception to science is delineated at the end of the examination of demonstration because demonstration departs initially from principles which cannot be demonstrated..." (p. 27).

[10] By emphasizing Topics I, 12 and the inductive mode of dialectic there, Mackeon ignores the mode of reasoning mentioned in I, 2 to which the mandate of establishing the archai of the various sciences is given (p. 28, note 63).

[11] Karl Popper, Conjectures and Refutations (New York: Harper & Row, 1963). "It can be shown that what is usually called the simplicity of a theory is associated with its logical improbability, and not with its probability, as has often been supposed. This, indeed, allows us to deduce from the theory of science outlined above, why it is always advantageous to try the simplest theories first. They are those which offer us the best chance to submit them to severe tests: The simpler theory has always a higher degree of testability than the more complicated one" (p. 61). The discussion that follows has been greatly informed by Popper's ideas on versimilitude. See his Conjectures and Refutations (pp. 228-248) and his essay on "Two Faces of Common Sense" in Objective Knowledge (Oxford: Clarendon Press, 1972), chapter 2, especially sections 6 through 11.

[12] Hempel, p. 7.

[13] Hempel, p. 7.

[14] See Popper's "My Solution of the Problem of Induction," chapter 1, in Objective Knowledge. Says Popper (and I am sympathetic): "My answer to the problem is the same as Hume's" No, we cannot; no number of true test statements would justify the claim that an explanatory universal theory is true" (p. 7).

[15] The contemporary information in this and the following paragraph was drawn from Isaac Asimov's The Genetic Code, chapter 1, "Inheritance and Chromosomes."

[16] See Objective Knowledge, pp. 119, 144, 164, and 287.

[17] See Erna Lesky's Die Zeugungs--und Vererbungslehren der Antike und ihr Nachwirken (Wiesbaden: Franz Steiner Verlag, 1950) and Konrad Blersch's Wesen und Entshehung der Sexus im Denken der Antike (Stuttgart-Berlin: W. Kohlhammer Verlag, 1937). Also, Hans Meyer's "Das Vererbungsproblem bei Aristoteles" in Philologus LXXV, 1919.

[18] "And one further relevant consideration is this: if more and more qualifying hypotheses have to be introduced to reconcile a certain basic conception with new evidence that becomes available, the resulting total system will eventually become so complex that it has to give way when a simple alternative conception is proposed"(p. 30).

[19] This is the title of chapter two of Goodman's Fact, Fiction, and Forecast (Indianapolis: The Bobbs-Merrill Co., Inc., 1965), "The dispositions or capacities of a thing--its flexibility, its inflammability, its solubility--are no less important to us than its overt behavior, but they strike us by comparison as rather ethereal. And so we are moved to inquire whether we can bring them down to earth; whether that is, we can explain disposition--terms without any reference to occult powers" (p. 40).

[20] R. Harré, "Powers," British Journal of the Philosophy of Science, 1970, 21, p. 83.

[21] Harré, p. 85.

[22] Harré, p. 86.

[23] Harré, p. 88.

[24] See Generation and Corruption II, 7, especially 333b23 and 334b14.

[25] Here I am indebted to Terry Penner. Also, see H.S. Rhayer's paper "Aristotle on Nature: A Study in the Relativity of Concepts and Procedures of Analysis," The Review of Metaphysics, June 1975, vol. XXVIII, no. 4.

[26] Harré, p. 90.

[27] Harré, p. 90.

[28] Harré, p. 91.

[29] J. L. Mackie, Truth, Probability and Paradox (Oxford: Clarendon Press, 1973), p. 143.

[30] For a discussion of this topic, see David Hull, Philosophy of Biological Science (Englewood Cliffs: Prentice-Hall Inc., 1974), I, "The Reduction of Mendelian to Molecular Genetics."

[31] Moliere (seventeenth century) is the first thinker to raise suspicions about the sleep-inducing power of opium. This (if true) is no surprise, for at that time scientists had started to penetrate deeper into the nature of things.

[32] Rudolph Carnap, "Empiricism, Semantics, and Ontology," reprinted in Semantics and the Philosophy of Language, edited by Leonard Linsky (Urbana: University of Illinois Press, 1952).

[33] Carnap, p. 211.

[34] Carnap, p. 211.

[35] Hans Driesch, The History and Theory of Vitalism, translated by C. K. Ogden (London: Macmillan, 1914), p.1.

[36] Hull, p. 129.

[37] Hempel, p. 72.

[38] See Hull, p. 92ff.

[39] Jacques Monod, Chance and Necessity, translated by A. Wainhouse (New York: Random House, 1972).

[40] Monod, p. 9

[41] See Physics II, 1.

[42] Monod, p. 12.

[43] The progress of science "required the unbending structure implicit in the postulate of objectivity--ironclad, pure, forever undemonstrable. For it is obviously impossible to imagine an experiment which could prove the nonexistence anywhere in nature of a purpose, of a pursued end" (p. 21).

[44] Monod, p. 24

[45] Monod, p. 28.

[46] Monod, p. 46

[47] Monod, p. 29.

[48] See Monod, Chapter VII.

[49] Monod, p. 118.

[50] Monod, p. 121.

[51] Monod, p. 82.

[52] Monod. p. 87.

[53] Donald Davidson, "Actions, Reasons, and Causes," The Journal of Philosophy, Vol. LX, no. 23, November 1963, p. 698.

[54] Hempel, p. 70.

[55] Mendel himself did not use the word "gene" in his famous report on his researches with peas. See Ernest B. Babcock, The Development of Fundamental Concepts in the Science of Genetics (Washington, D. C.: American Genetic Association, 1950), p. 9.

[56] This is the title of chapter XXVII of Darwin's The Variation of Animals and Plants Under Domestication (New York: Orange Judd & Company, 1868), vol. II.

[57] Darwin, pp. 481 and 482.

[58] See Hull, chapter 1.

CONCLUSION

Let me briefly enumerate the theses I have and have not defended.

1. In the last chapter, I did not argue that Aristotle's explanation of the generation and development of animals was the correct explanation or even the best one so far. This ultimately is a matter for present-day biochemists and geneticists to decide. We may note, though, that even an incorrect explanation is still an explanation and can be evaluated as such.[1] In fact, I compared and contrasted two theories, both of which do not seem to reflect current thinking to any accurate degree.

2. Instead, I have taken an episode out of the history of science and philosophy, placed two competing theories in their contexts, analyzed and evaluated them, and decided that Aristotle's form-matter hypothesis was a step forward in the history of science.

3. More specifically, I argued that throughout the G. of A., Aristotle is dialectically engaged with the Hippocratic author of On Semen and On the Nature of the Child. This is the historical component of my thesis. I have shown that Aristotle wrote large and important passages of the G. of A. with the contents and arguments of these two treatises in mind.

4. Also in Chapter II, I argued the philosophical side of the thesis by defending the claim that the dispute between Aristotle and our Hippocratic author is indeed a dialectical dispute in the proper sense of the word. In the second chapter of the Topics, Aristotle says that dialectic has the task of establishing the fundamental principles or archai of a science. This is precisely what is going on in Aristotle's dispute with the Hippocratic author. The two men disagree on the archai of biology. Our author, who was a defender of the pangenesis theory, believed that the sexes make similar contributions, while Aristotle proposed the hypothesis that the male supplies the form and the female the matter. We must, argues Aristotle in Book I, accept either the pangenesis theory or his own hypothesis. He is quick to show, however, that the pangenesis theory is totally defective. It leaves us with the problems of development, resemblance, and sex differentiation. I discussed these problems in Chapter III.

5. I then showed in Chapters IV and V that the dialectical dispute is carried over into Book II of the _G. of A._ and is brought to a conclusion in Book IV. In the former, Aristotle gives his own solution to the problem of development, while in the latter he tackles the problems of resemblance and of sex differentiation, offering quite a sophisticated theory of inheritance. This theory explains how any case of resemblance deviates from the ideal case, which is a boy who looks like his father. Not until Aristotle has explained all the various cases of resemblance can he be said to have met the strongest evidence on behalf of (P). The discussions of Book IV are, therefore, a necessary complement to those of Book I, where Aristotle did not explain how a girl could look like her mother and other deviations from the ideal.

6. We are now in a position to realize that the form-matter hypothesis is not a dogmatic hangover of Aristotle's platonic days. He did not blindly impose an utterly foreign distinction on recalcitrant biological data. Rather, he proposed this hypothesis as a genuine scientific hypothesis designed to meet certain pressing problems in biology. And he did an excellent job in making his proposed solution stick. Furthermore, he viewed both the raising of objections to (P) and the defense of his own hypothesis as the proper task of the science of dialectic.

7. There is, therefore, something wrong with the prevailing negative opinion, according to which science and dialectic are not very closely connected for Aristotle. Not only does Aristotle explicitly link the two disciplines in _Topics_ I, chapter two, he also practices the mandate of the Topics a great deal. The present study is a case in point. Aristotle practices in the _G. of A._ what he preaches in the _Topics_.

8. This calls for a reevaluation of Aristotle's views on the nature and procedures of science in the _Posterior Analytics_. Either we are left with a clear difference on a point of method (since the _Analytics_ bars opinions from playing a role, whereas the _Topics_ gives them a crucial function) or we must seek to bring the preaching of the _Analytics_ in line both with Aristotle's practice and with his preaching in the _Topics_. It may well be that Aristotle is much less of a Baconian than we have taken him to be. Certainly, what emerges from this study is a delicate picture of travelling from the theories to the facts, and back again to the theories until the outstanding problems are solved.

9. This difference on method is not removed by Owen's suggestion that ultimately the results of both methods are to be tested against experience. With this I agree. In fact, I have gone even further than Owen in this respect. He seems to think that dialectic has as task the establishment of *a priori* categories and analytic truths. But he ignores the empirical aspect of even that enterprise. For as I showed, even in the discussions of incontinence and of place, which for the moment could be labeled *a priori* categories, Aristotle does have clear and direct recourse to experience. Then, too, this study shows--and the mandate of the *Topics* was intended this way--that it is the task of dialectic to establish a straightforward empirical hypothesis. For this is how I see the form-matter hypothesis; it is not an *a priori* category, but a genuine scientific hypothesis designed to solve certain very specific and concrete problems in the philosophy of nature. In other words, it is not true to say that Aristotle practiced the Baconian method in his biological treatises. At least it does not apply to the one analyzed in this study.

10. It turns out then that Aristotle was a more sophisticated scientist than we sometimes think he is. He did not come to the data with all sorts of ill-conceived and preconceived distinctions which he then forced upon them. He did not believe that scientific theories could be derived from the facts in a simple, inductive way. Rather, for him, scientific progress was a matter of making conjectures and refutations. The form-matter hypothesis, the potential-actual distinction, his vitalism, and his theory of inheritance all fit into this merger of science and dialectic.

The implication of this study is that Aristotle's biology is not as sexist as is sometimes thought. I have placed his form-matter application in its historical and intellectual context and shown that Aristotle offered it as a scientific hypothesis. When seen in its context, the theory testifies to Aristotle's scientific acumen, and it would seem that in constructing it, Aristotle stayed within the boundaries of accepted scientific practice.

INDEX

Achinstein, Peter, 177 note 1.
Aeschylus' Eumenides, 51.
Analytic-Synthetic, 12 ff.
Anaxagoras' belief that in everything there is a portion of everything, 61 note 12, 78.
a priori, 15, 22.
archai, 1ff, 43ff.
art-nature analogy, 53f, 94f, 104, 107f, 162, 165.
Aristotle's reputation as a scientist, 50f.
Asimov, Isaac, 173 note 15.
Aubenque, P., 60 note 3.
Automota, 97, 122 note 7.

Babcock, Ernest B., 176 note 55.
Baconian, 2, 13, 15, 26, 149.
Balme, D. M., vii note 1, 64 note 43, 75, 84 note 1, 85 note 7.
Barnes, J., 35 note 55.
Blersch, Konrad, 174 note 17.
Bos, A. P., 123 note 22.
Brody, B., 51, 63 note 32.

Charlton, W., 101ff, 122 note 11, note 12.
Cline, Horowitz, Maryanne, vii note 6.
Carnap, 159, 175 note 32.
contact between agent and patient, 95ff.

Darwin's pangenesis theory, 61 note 9, 167, 176 note 56.
Davidson, Donald, 166, 176 note 53.
Democritus, 46f, 150, 164.
development, problem of, 94ff.
dialectical discussion, features of, 48ff.
dialectical character of research, 5ff, 37ff.
dialectic, nature of, 42f, 60 note 4, 73.
dispositional properties, 155.
dominance principle, 136ff.
Driesch, Hans, 175 note 35.
dunameis, as powers, 156
as theoretical entities, 166
dunamis, meaning of, 132
Düring, I., 31 note 8, 52, 64 note 36, 142, 145, note 10.

Empedocles, 80, 157, 164.
endoxa, 12 ff.
epigenesis, 51.
eristic arguments, 65 note 45, 172 note 2.
evolution, theory of, 172 note 4.

female as archē of biology, 43ff.
female does not secrete semen, 56.
form-matter distinction, used as hypothesis, 50ff, 94ff.
content of, 151.
scope of, 147ff.
combined with potential-actual into one principle of explanation, 99, 122 note 8.
formal cause of procreation is the male parent, 106ff.

Goodman, N., 152, 156, 159, 170 note 19, 174 note 19.
Grene, M., 2, 3, 51, 63 note 31, 125 note 30.
growth, 78ff.

Harman, Gilbert, 172 note 6.
Harré, R., 156, 174 note 20
Hempel, C., 161, 166, 172 note 7.
hylomorphism compared with pangenesis, 43ff.
hypothesis, 50ff.
Happ, H., 123 note 21.
Hippocratic author, on embryonic development, 37, 45ff, 67ff, 89ff, 160.
Hippocratic treatises, unity of, 121 note 1.
Hull, D., 161, 175 note 30, 176 note 58.

induction, 75ff, 150.
inheritance, theory of, 131ff.
incontinence, discussion of NE VII, 3, 19ff.
invariance, 162ff.

Jaeger, Werner, 31 note 8.
Jones, Barrington, 101ff.

Kahlenberg, W., 121 note 1.
Kember, O., 61 note 12.

Le Blond, J. M., 5, 30, 32 note 15.
Lewes, G. H., 2, 31 note 1.
Lewis, C. I., 17, 34 note 42.
Lesher, J. H., 6, 32 note 16.
Lesky, E., 46, 62 note 13, 84 note 1, 142, 144 note 3, note 4, 145 note 8, 174 note 17.
like goes to like, used as explanatory maxim, 87.
replaced by the potential-actual distinction, 92ff.
Littré, 62 note 14.
Lloyd, G. E. R., 64 note 37, 31 note 8.
Lones, T. E., 31 note 8., 49, 63 note 28.

Mackeon, Richard, 150, 172 notes 8-10.
Mackie, J. L., 157f, 175 note 29.
male as archē of biology, 43ff.
male does not contribute matter, 151.

Mansion, S., 39, 60 note 2.
material cause of procreation is the menstrual blood, 100ff.
Mendel, 158, 176 note 55.
Meyer, Hans, 135, 144 note 2, 174 note 17.
Michael of Ephesus,
Mill, J. S., on hypothesis, 53, 64 note 38.
Miller, H., 144 note 1.
Mohere, 175 note 31.
Monod, Jacques, 158ff, 175 note 39, note 43.
Moraux, P., 66 note 45.
multiple births, 82, 104.
mutilations, argument from, 70, 104, 141.

Needham, J., 31 note 8, 50f, 55, 63 note 26, 65 note 44, 141, 144 note 7.
Nuyens, F., 124 note 29.

On the Nature of the Child, compared to G. of A., Book Two, 93, 121 note 3.
On Semen, compared to G. of A., Book One, 45f.
Owen, G. E. L., on characteristics of dialectic, vii, 12ff, 42, 60 note 4, 62 note 19.
oviparous fish, 57.

pangenesis, arguments for, 67ff.
  arguments against, 73ff.
  compared with hylomorphism, 43ff.
  used to explain resemblances, 132ff.
parts of animals, 74.
Peck, A.L., 44, 60 note 8, 62 note 15.
potential-actual distinction, introduced, 87ff.
  charge of vacuity against, 155ff.
  used to explain development, 94ff.
Peirce, Charles, 149, 172 note 5.
Penner, T., 35 note 43, 66 note 45, 123 note 24.
phainomena, 12 ff.
Physics, place, discussion of, 16, 26ff.
Platt, 141, 144 note 6.
pleasure of copulation, 68f.
pneuma, 111ff.
Popper Karl, 52, 64 note 40, 149ff, 173 note 11, note 14.
Posterior Analytics II, 19, 5ff.
powers, 156ff.
Preus, Anthony, v, vii note 5.
Putnam, H., 17, 34 note 41.

Quine, W. V. O., 16, 33 note 40.

Randall, J. M. Q., 3, 31 note 9.
Regis, L. M., 4, 31 note 10.

relapse principle, 136ff.
resemblance, argument from, 70f, 127ff.
Rhayer, H.S., 174 note 25.
Ross, D. W., 9ff, 19, 32 note 17, 63 note 30, 73, 85 note 5.
Ryle, G., 62 note 18.

semen, as concocted blood, 129
as residue of nourishment, 55f.
nature of, 45ff.
scientific
theory, 166
revolution, 87, 93
growth, 149.
method, 5ff.
sex differentiation, problem of, 80ff.
in reproductive process, 80, 94
Singer, Charles, 31 note 8, 50, 51, 63 note 25, 64 note 34, 141, 144 note 5.
Socrates, on weakness of will, 20ff.
Solmsen, Friedrich, 4, 32 note 13.
Souilhé, J., 144 note 1.
soul, 44
acquisition of, 111ff.

teleonomy, 162.
Theophrastus, 52, 75.
Thompson, D'Ary, 31 note 8.
Topics I, 5ff, 37ff.
transplants, 75.

Veatch, H.B., 124 note 28.
vitalism, 161ff.

wind eggs, 57
Weil, Z., 35 note 55.